HOUSING AND THE SPATIAL STRUCTURE OF THE CITY

Residential mobility and the housing market in an English city since the Industrial Revolution

R. M. Pritchard

CAMBRIDGE UNIVERSITY PRESS

CAMBRIDGE

LONDON · NEW YORK · MELBOURNE

Published by the Syndics of the Cambridge University Press
The Pitt Building, Trumpington Street, Cambridge CB2 1RP
Bentley House, 200 Euston Road, London NW1 2DB
32 East 57th Street, New York, NY 10022, USA
296 Beaconsfield Parade, Middle Park, Melbourne 3206, Australia

© Cambridge University Press 1976

Library of Congress cataloging in Publication Data
Pritchard, Roger Martin, 1946–
Housing and the spatial structure of the city.
(Cambridge geographical studies ; 7)
Bibliography: p.
Includes index.
1. Residential mobility–Leicester, Eng.–History. 2. Housing–Leicester, Eng.
3. Leicester, Eng.–Social conditions. I. Title. II. Series.
HD7334.L43P74   301.32'6'094254   75–3859
ISBN 0 521 20882 3

First published 1976

Printed in Great Britain by
Cox & Wyman Ltd,
London, Fakenham and Reading

CAMBRIDGE GEOGRAPHICAL STUDIES

Editorial Board: B. H. FARMER, A. T. GROVE, E. A. WRIGLEY,
B. T. ROBSON

7 HOUSING AND THE SPATIAL STRUCTURE
OF THE CITY

CAMBRIDGE GEOGRAPHICAL STUDIES

CONTENTS

Contents

LIST OF FIGURES AND TABLES

List of figures and tables

ACKNOWLEDGEMENTS

Anyone who is vain or foolhardy enough to embark on a large piece of research, especially when it takes him beyond the confines of his original training, accumulates debts, the acknowledgement of which is poor repayment for all they have meant in help and encouragement.

Firstly, I must thank all those persons and organisations in Leicester who gave so much aid and hospitality, and whose co-operation provided the bulk of the material on which this study is based. They are too numerous to mention everyone by name, but special thanks must be given to the Registrar-General and the staff of the Leicester Registration district for allowing use to be made of marriage and death registers; to the staff of the Archives Department of the Leicester City Museum (now incorporated into the Leicestershire County Records Office); to the City Treasurer; the City Planning Officer; the City Librarian and all their staffs.

Also I must thank all my former colleagues in the Geography Department of the University of Cambridge, and especially my supervisor, Dr Brian Robson of Fitzwilliam College, from whose perceptive comment and criticism I have benefited enormously. A special debt is also due to all my present friends and colleagues in Bristol.

But most of all thanks must go to my parents.

CHAPTER 1

INTRODUCTION

This is a study of the relationship between housing and the geography of the city examined through the changing pattern of residential mobility. The manner and frequency with which people change their homes is obviously a process deeply rooted in the spatial organisation of urban areas, but specifically spatial factors cannot explain its character. Although such conditions as proximity or direction are significant parameters in any equation of movement, they can only be made meaningful if considered in context and that context of residential mobility is necessarily the total social, political and economic condition of the city. The resulting limitations for any study in time and expertise are obvious. The solution adopted here has been to frame this investigation within the structure and operations of the urban housing market. Indeed, the more deeply I looked into these problems, so the significance of housing for the spatial organisation of the city loomed larger and larger.

In the most direct sense, housing must be an important component of urban structure if only because it is the predominant land use. It is a truism that cities are neither mere physical artefacts nor simply abstract concepts. Housing, through the role of politics and public policy, is one of the key elements which link both material and social aspects of the city. It is also a topic that necessarily straddles the boundaries of those traditional, academic disciplines which have interested themselves in urban problems. Today, such are the complexities of housing – economic, social, political and legal – that no single subject can possibly claim a monopoly of knowledge. It is surely no accident that so much of the pioneering work of recent years should have come from the like of Donnison and Cullingworth who are concerned with essentially 'practical' topics such as social administration and town planning.

The environment of the city is never static and the relationship between housing, migration and urban spatial organisation has to be reinterpreted continually in terms of shifting social, economic or political circumstance. Here, the changing patterns of residential mobility and intra-urban migration are seen as providing a fresh insight into the manner in which the geography of British cities has developed as a reflection of these social, economic and political forces. Such interactions are obviously long-term processes evolving slowly through time, yet to study residential mobility in a historical context is no easy task. Information is scarce and often of dubious quality, a fact reflected in the dearth of previous studies.

Because housing has so central a role in the city, no study can hope to do justice to all its facets. I cannot pretend to have dealt adequately with every aspect of residential mobility, let alone the complete range of influences which the housing market has upon urban forms and processes. What this study tries to illustrate, using the example of one English city, Leicester, since the Industrial Revolution, is the complex character of these relationships which bind together such large-scale forces in urban society with an individual's residential location decision, and how this is then reflected in the ever-changing social geography of the city.

Although I have taken what is essentially a geographical or spatial viewpoint of such topics, their nature has constantly brought me into contact with those from other disciplines, who starting from different premises frequently seem to find themselves concentrating on the same fundamental issues. It is often bemoaned that the enormously involved ramifications of the city, which cause everything to be so intimately linked with everything else, make it extremely difficult to analyse. But those linkages ought to be a continual reminder of the limitations imposed by the intellectual frameworks that are our individual academic disciplines. In the city, one may take an economic, or a sociological, or a geographical approach, but we all end up looking at urban problems. This is a viewpoint which has been brought home to me most strongly since leaving academic life for the day to day tasks of urban and regional planning, where the questions faced have set so many of my concerns here in a new and refreshing perspective.

CHAPTER 2

HOUSING AND THE SPATIAL STRUCTURE OF THE CITY

Housing would take a central place in geographical studies of urban areas if for no other reason than its physical significance. Residential land accounts for about half the total area of our towns and cities (Best & Coppock, 1962, p. 38). Of course, every urban area has many other land uses – industry, recreational open space, commercial properties, schools and colleges – but residential usage is the largest single category, and the most general. Every town or city specialises in certain social or economic functions (Moser & Scott, 1961) which are reflected in the make-up of *their* land-use patterns, but all must devote a large proportion of their total area to housing.

At the same time, everyone requires shelter, a roof over his head, and for most of us that means a home, a permanent 'base' in which we spend the greater part of our time. Naturally, housing is an important factor in most peoples' lives and this importance is reflected in its social and political significance. Housing is a major plank in every party's platform at all elections, and there is a long history of lobbyists and pressure groups who have seen housing deficiencies at the core of society's problems.

Because of this obvious physical and human significance, housing must relate the built-form of an urban area to its social structure. As Harvey has pointed out, these two aspects of the city 'should be regarded as complementary. The trouble is that the use of one impedes the use of the other. Any successful strategy must appreciate that spatial form and social processes are different ways of thinking about the same thing.' (Harvey, 1970, p. 48). The strength of this relationship in the case of housing is especially highlighted both in the way in which new dwellings are built and occupied and in the manner in which the housing market responds to long-term changes in the urban environment.

Very few people have complete freedom to choose where they live. Individual residential location decisions are not taken in a vacuum. Very few families have any great hand in supervising the construction of the house which they are to occupy. The overwhelming majority select from that limited range of suitable properties, available at the time, which best meets their particular means and needs. For some, the range of dwellings may be a wide one; for others, especially those on low incomes, it may be very narrow. Every decision will involve the resolution of choice and constraint, and for all, the limits of that decision will be set by the actions of those responsible for the provision of accommodation.

3

For example, if one considers only those responsible for new building today, they consist of a great many agencies, groups and individuals operating on a number of levels within a housing market of enormous, and growing, complexity. No longer are new houses simply the responsibility of the individual builder. Construction is now the operational end of a long chain of financial and administrative controls. This stretches from the implications of national economic policy, as set by the Government through the Treasury, down to the institutions who provide the direct capital for new building – banks, insurance companies, building societies – and the national and local authorities who provide much of the infrastructure for development as well as setting the framework of planning decisions which may determine the basic distribution of land uses, before it reaches those who actually build houses.

Furthermore, today there are many such housebuilding agencies; not only the traditional, small, speculative builder, but also large commercial development firms who have branched out into residential construction on a national scale;[1] local authorities, who now provide between a third and a half of all new dwellings; and numerous specialist agencies such as housing associations and charitable trusts. With such a system, the current housing market is not just a marriage of individual social circumstances to local physical constraints, but a result of many of the most fundamental economic and social forces at work in Britain today.

It is also important to recognise the current state of housebuilding as just a snapshot of one section of the long-term processes and relationships which exist in the housing market. The proportion of the housing stock replaced or augmented in Britain today only accounts for about 2 per cent of total dwellings. With a high capital cost, housing is naturally distinguished by its durability. The life-span of the average house in present British cities is about a hundred years. Table 2.1 shows the current (1971) age-distribution of

TABLE 2.1   *The age-distribution of dwellings*

Age	% of Stock
Built before 1891	20
Built 1892–1918	15
Built 1919–44	25
Built 1945–71	40

dwellings nationally (Riley, 1973). A third of the present housing stock dates from before the First World War, and therefore was produced by a system of development very different from that in existence today. The most obvious example of the changes that have occurred is the alteration in the tenurial conditions under which property is occupied. In 1914, about 95 per cent of all houses were let by private landlords; by 1971, this had fallen to about 15 per cent. Owner-occupation, which accounted for only about 5 per cent of the market in 1914, now takes up half, and letting by local authorities, which was insignificant before the First World War, has risen to 28 per cent of all dwell-

ings. Tenurial changes are just the most obvious symptoms of the fundamental alterations that have come about since the turn of the century, when
so many of the features that are today taken for granted – subsidies, rent
controls, security of tenure, and especially the whole range of town planning
legislation – did not exist (Cullingworth, 1966; 1970; Donnison, 1967;
Cherry, 1974).

Every contemporary residential location decision takes place within the
context of large numbers of past events, which though they may have occurred
originally in a very different environment, still produce consequences which
influence behaviour in today's cities. Such a matrix of historic constraints
operates in all facets of urban life, but the character of housing makes it
especially important in this field. Necessarily, such a relationship is a very
important factor in determining the direction and speed with which any city's
spatial organisation is adjusted and adapted to meet the strains of growth,
change or decay. It also implies that such adjustments and adaptations will
involve a very complex process. It has already been suggested that every city
results from the marriage of a foundation of legislation, economic policy and
social structure which might be common to a nation or even to a culture, to
thousands (if not millions) of individual decisions, taken in the particular
context of an ever-changing local environment. Such a spatial hierarchy of
decisions is paralleled by a temporal hierarchy resulting from the differential
leads and lags produced by all the many separate elements in the situation.

A very good example of this is the long period (perhaps decades) required
to accomplish many of the consequences of some change in the legislative
framework, which may need such thousands of individual decisions to
produce all the circumstances, many of them originally unforeseen, that may
result from its interaction with the many fluid variables significant in the
housing market.

It is mainly with these interactions and their effects on the spatial organisation of the city that this study is concerned. Such complications not only
emphasise the difficulties involved in reconciling the temporal and spatial
dimensions of housing but equally demonstrate the impossibility of adopting
a blinkered approach to the spatial structure of the city as interpreted through
its housing market. Such is housing's central place in urban life that it must be
examined from many viewpoints, and consideration ought to be given to the
way the major urban social sciences have approached the problem.

Analysis of the housing market

Though housing has been a major subject of social significance and political
controversy at least since the Industrial Revolution, any comprehensive
analysis of its structure and workings has been very restricted by a scarcity of
basic information. The strongest line of housing research has always been that
one which derives from the nineteenth century concern with urban public

health (Ashworth, 1954). There is a long history of Parliamentary Enquiries, Royal Commissions and all forms of official investigation which have been especially influential in shaping public attitudes and actions towards housing. Such concerns have been reflected in the sort of statistical basis provided by the Census of Population, which until very recently concentrated exclusively on physical measurements of house condition, such as sanitary facilities and overcrowding.[2] As a consequence, far more is known about the structural quality of the dwelling stock and the way in which this has changed and been improved than of the manner in which the housing market operates or responds to changes in the general social and economic climate (Duncan, 1971).

Today, the situation is perhaps improved, thanks to a considerable combined effort by both public and private research. The Government Social Survey led the way with the National Housing Surveys of 1960 and 1964 (Gray & Russell, 1962; Woolf, 1967), and contemporary with these were the Rowntree Trust's reports on the workings of the 1957 Rent Act and current housing trends (Donnison, 1961; Cullingworth, 1965). National research also stimulated investigations at the regional and local levels. Cullingworth's survey of Scottish housing in 1965 (Cullingworth, 1967) has been followed by an on-going series of Government-sponsored research reports for areas of 'housing stress' such as the major conurbations (West Midlands Conurbation Housing Survey, 1971). Publically available local surveys are rarer (though see Glass, 1969; Karn, 1970), but many unpublished studies have now been carried out in connection with local authorities' statutory responsibilities towards slum clearance, General Improvement Areas and council housing.[3]

Such surveys usually take a similar form. They set out the basic housing conditions in an area – tenure, age of property, standards of facilities and maintenance, costs and values – and relate them to the character of the area's inhabitants – age, income, social class, size of household, length of residence and so forth. Sometimes, they have also included sections on the inhabitants' satisfaction with their dwellings, any recent moves made by resident households and any desire or intention to move in the future. They have created a situation where much more is known about the elements which make up the present housing market in Britain, and something about the current patterns of behaviour that determine its operations.

Yet accumulating evidence has demonstrated the complexity of the contemporary British housing market and has reinforced the opinion that it is composed of a number of sectors, especially associated with the differing tenures under which property is held, the interactions between which remain only imperfectly understood. It is difficult, if not impossible, to articulate accurately the manner in which changes in one sector of the market may work themselves through to all the others. As has been said in conclusion to one of the most comprehensive of such housing studies, such a survey, 'can

6

merely sketch the main features of the housing situation at a moment in time; it can do little to unravel the mechanisms of the housing market or the factors shaping households' attitudes and preferences' (West Midlands Conurbation Housing Survey, 1971, p. 73). Such problems are reflected in the considerably less attention devoted to longer-term processes in the housing market. Though there are a large number of studies which have examined specific aspects of housing policy or accommodation problems (e.g. Sutcliffe, 1974), general considerations of the long-term operations of the housing market in any particular local context are much rarer (but see Cullingworth, 1963, for a study of Lancaster). Consequently, any attempt to form a framework within which the relationship between housing and the spatial organisation of the city might be examined, has to be derived from looking at the differing standpoints from which the main urban social sciences have viewed such problems. In particular, it is worthwhile investigating the way in which previous geographical examinations of the evolution of British cities have treated housing and its role, and then comparing how economists and sociologists have approached the same subject to see what could be usefully learnt for incorporation into a geographical analysis.

Geographical approaches to the evolution of British cities

Studies of the way in which the present distributions of people, buildings and functions in cities have come about are a central theme for urban geographers. Growth, change and decay are the commonplaces of the urban condition, and with very few exceptions, towns and cities do not suddenly appear in any complete or final form, or usually ever achieve such a state.[4] In the most simple terms, there have been two main approaches in the geographical study of the evolution of urban areas in Britain – the morphological and the ecological.

The morphological approach grew out of an older and more general concern of geographers with the 'landscape' and its organisation into that distinctive spatial unit, the 'geographic region' (Dickinson, 1947). At first, towns and cities were commonly seen as one of many features in the regional landscape, but there soon developed a specific interest in the internal structure of urban areas themselves.

Smailes summarised those elements which he felt should compose any geographer's study of town or city,

differences in either or both these intimately related aspects of urban morphology, function and form, give a basis for the recognition of urban regions. It is the description of their nature, their relative disposition, and their social interdependence that constitutes a geographical analysis of an urban area (Smailes, 1953, p. 84.)

This approach was especially influential in the years immediately after the Second World War when a number of studies were carried out which, in their search for spatial organisation in the morphology of the city, said a great deal about the historical context in which urban areas in Britain had evolved (Haughton, 1949; Rodgers, 1952; Thurston, 1953; Smailes, 1955).

Morphological analysis remains an important tool in urban geography. Carter's *The Towns of Wales* illustrates how such an approach may form the base for the comparison of a number of urban areas (Carter, 1968), whilst there are also a number of recent studies which examine individual, structural elements within the city's fabric (e.g. Whitehand, 1967; Johnston, 1969a). The most important contribution of morphological analysis has been its emphasis on the continuity and temporal interdependence of urban areas. This is excellently demonstrated by Conzen's study of the development of the small Northumberland town of Alnwick (Conzen, 1960). Here, limited scale allowed the evolution of the present distribution of buildings and land use in the town to be reconstructed plot by plot, property by property, like a giant jigsaw puzzle.[5]

But there have been criticisms and objections to the morphological approach. It is a methodology which seems to work best at a smaller scale, like the Alnwick study, and it has not always produced meaningful general statements which have gone much above the most cursory description when faced with larger areas. Most importantly, a morphological analysis of urban evolution has been criticised for failing to pay sufficient attention to the social processes which lie behind spatial forms. Jones stated very succinctly that, 'To recognise the city as a product of social forces is not enough without also studying its society.' (Jones, 1966, p. 11.) This limitation is well illustrated by the manner in which morphological studies have treated housing. In delimiting urban regions, major characteristics of house type – age, size and condition – have often been used as evidence. Thurston's study of St Albans is a good example (Thurston, 1953) and morphological work in recent years has placed great emphasis on the examination of house-types in the building fabric (Whitehand, 1965). Yet because of the only indirect acknowledgement of the role of social processes in the development of urban spatial structure given by morphological analysis, it is not surprising that many geographers should have adopted the alternative framework provided by the ecological analysis of urban areas.

Human ecology has been enormously influential in urban studies, but it is now very difficult to summarise adequately all the viewpoints that have been expressed under its banner (Hatt & Reiss, 1957; Theodorson, 1961; Burgess & Bogue, 1964). 'The ecological perspective is so loosely defined that it can be stretched to include what is praiseworthy . . . and contracted to avoid the apparently faulty.' (Rossi, 1959, p. 148; quoted by Robson, 1969, p. 25.) The ecologists whose ideas were formulated at the University of Chicago in the 1920s were greatly influenced by those processes which they observed going on in American cities at the end of the nineteenth and beginning of the twentieth centuries. They saw how immigrant groups, recruited from the peasantries of Ireland, Central and Eastern Europe (and latterly from the Negro communities of the Southern States) faced the new conditions of urban, industrial life and the consequent necessity to reconcile the contradictory demands of their traditional cultures with the needs of their new environment.[6]

The ecologists thought this conflict implied an essentially pathological character to urban life and their research topics reflected this (e.g. Thrasher, 1927; Wirth, 1928; Zorbaugh, 1929). Equally, such conflict had a necessary corollary in the formation of new, urban, social patterns needed for the acculturation of such immigrants into the life of the city,

Disorganisation as a preliminary to the reorganisation of attitudes and conflict is almost invariably the lot of the newcomer to the city, and the discarding of the habitual, and often what has been to him the moral, is not infrequently accompanied by sharp mental conflict and sense of personal loss. (Burgess, in Burgess, Park & McKenzie, 1925, ch. ii.)

Inevitably, the more mature cities of the past fifty years have brought forth sociological criticisms of the dominant, simplistic themes of early, ecological work (e.g. Martindale, 1958; Beshers, 1962; Reissmann, 1964; Sjoberg, 1965). Ecologists themselves have been well aware of such criticisms and Rossi's previously quoted view of the loose definition of human ecology stems in large part from the continuing reformulations that have taken place (e.g. Hawley, 1950), but recent attempts to establish the soundness of the theory's sociological pedigree (Schnore, 1958; Schnore & Duncan, 1959) have not lessened this criticism.[7]

By and large, such arguments have remained outside the compass of geographers' interests in the theory of human ecology. Its continuing significance for urban geographers has lain in the simple fact that the ecologists themselves recognised a key role for the spatial organisation of the city. Park, one of human ecology's original proponents, wrote in 1925,

It is because social relations are so frequently and so inevitably correlated with spatial relations; because physical distances so frequently are, or seem to be, the indices of social distances, that statistics have any significance whatsoever for sociology. And this is true finally because it is only as social and physical facts can be reduced to, or correlated with, spatial facts that they can be measured at all. (Park, 1925, p. 18.)

Burgess's ideal type of city structure, the concentric ring model, was the logical derivative of this principle and of the behaviour that the ecologists expected. (For the Burgess model, see Timms, 1971, pp. 211–23.)

It has been Burgess's model and the numerous modifications that have stemmed from it that geographers have found most useful (Robson, 1969; Timms, 1971; Johnston, 1971). At best, this has greatly increased knowledge of the relationship of the social and spatial structures of the city; at worst, it has degenerated into a fruitless search for geometric patterns based on a naive isomorphism.

Britain has never had the same range and availability of data as that upon which so much of the socio-geographic work had been based in the United States. The first study to use Census tabulations for small areas, the basis for so many American studies, was Collison's work on Oxford, which used specially processed information from the 1951 Census (Collison, 1959), but it was the general production of enumeration district statistics from the 1961 Census that triggered off a number of studies replicating the aims and methodology of earlier American research (Herbert, 1968; Robson, 1968; 1969).

Though such studies have been mainly concerned with contemporary circumstances, they have involved considerations of the way in which present social patterns have evolved (Robson, 1966; Gordon, 1971), and both contemporary and evolutionary studies have found that whilst there are similarities between American and British experience, there is not a direct correspondence. It is especially interesting that many of the differences seem to stem from, or are most strongly influenced by, differences in housing structure. Many of the earlier, ecological studies in Britain actually used Rating Valuations as a surrogate for social class (Jones, 1962; Timms, 1962; Herbert & Williams, 1962; Gittus, 1964–5; Robson, 1966; Gordon, 1971) and it has been shown how changes in the provision and organisation of housing have played a great part in re-orientating the spatial structure of British cities. Robson suggested this point in discussing the poor efficiency of the classical, ecological models in explaining the social geography of modern Sunderland,

The development of local authority housing both in peripheral areas and central re-development areas, together with the general increase in the role of central and local government planning, the spread of affluence, and its concomitant eroding of social differentials, or at least the increase in social and geographic mobility that this has brought with it, and the very fact of the great decrease in the rates of urban growth all have led increasingly to shortcomings in the classical models and to their limited usefulness in the modern setting. (Robson, 1969, p. 132.)

Certainly, this concern in Britain with the influence of housing on the spatial structure of the city has to be contrasted with the ecologists in the United States, where although it was recognised that housing had strong links with the city's social ecology, the mechanisms through which this was achieved and the consequent significance of changes in the organisation of the housing market were not directly examined.

It would be inaccurate to present the morphological and ecological approaches as rivals: rather they are complementary ways of looking at the same general problem of the evolution of the city's social and spatial structures. This complementarity is reflected in their treatment of housing. Though neither directly concerned itself with the workings of the housing market, nor even with the direct processes by which housing influenced and reacted with the spatial structure, both by their nature and interests had a great deal to provide by way of insights into such behaviour.

For the morphologists, concern with the built-form of the city and the functional distribution of land uses has had two major lines of influence. One has been on the development of research into urban history. Urban geographers' long-term concerns with the manner in which present urban forms represent a marriage of successive layers of development have brought them into close contact with the other disciplines which are combining with historians themselves to trace the growth of the city and its institutions, forms and structure (Dyos, 1968). The second influence has been through the growing profession of town planning, where the 1947 Town and Country Planning Act created a demand for people capable of drawing up land-use surveys and plans.[8]

For the ecologists, such aspects of physical development were subordinated to an interest in the manner in which the spatial arrangement of population sub-groups responded to such changes in the physical character of areas. The particular concerns of ecologists with ethnic and social segregation involved them much more closely with the processes of adaptation and adjustment and thereby with the interaction of social and spatial changes in the context of alterations in the quality of the housing stock.

What has been common to both morphological and ecological approaches has been that they have encouraged work which has almost always been at a high order of generalisation, dependent on inductive arguments capable of incorporating large-scale conceptions of urban structure intended to have wide applicability both spatially and temporally. This contrasts them with the deductive pattern of argument which has been typical of recent developments.

The strides made in constructing models of urban, spatial behaviour in the last few years have been enormous. From a concern with very limited, partial models concentrating on particular aspects of processes (e.g. shopping, journey-to-work, residential location), they have now progressed to the situation where it is possible to foresee the construction of general families of models capable of simulating a wide range of processes and conditions (Wilson, 1974, pp. 23–8, 261–8). The question of scale, both in time and space, remains of great importance. As Wilson (1974, p. 391) points out, many of the major difficulties faced in model-building derive from the mechanisms of change and he is especially concerned with the linking of processes and models developed at the micro-scale, e.g. the household or the firm, with that family of spatial interaction models which are being developed at the 'comprehensive' scale. But there is also a need to strengthen the links between this supposedly comprehensive level and the even higher scale of analysis represented by the inductive generalisations of morphological or ecological studies. The latter still form,

a foundation on which to build models that attempt to incorporate 'how' and 'why' residential location patterns come about . . . This often involves the model builder in a search for new or different concepts, different theoretical styles and different model building techniques. (Senior, 1973, p. 177.)

Obviously, the search for such links between these different levels will be a very difficult task, given the fundamentally different approaches that they have adopted.

One useful way in which the operational constraints of such links may be highlighted is through looking more directly at the methods both economists and sociologists have used in considering the housing market and more especially the long-term mechanisms of adjustment and adaptation which are necessary to reconcile the market's structure with changes in urban spatial organisation. In choosing such an approach, one is emphasising a very simple generalisation about the character of the two disciplines. This is that economics tends towards an aggregative structure of argument, whilst sociology adopts a disaggregative. Thus, the economist is most likely to build up larger-scale models from statements about individuals expressed in a theory of consumer behaviour or a theory of the firm (Wilson, 1974, p. 192), but the sociologist considers that,

> to understand the changes of many personal milieux, we are required to look beyond them . . . To be aware of the idea of social structure and to use it with sensibility is to be capable of tracing such linkages among a great variety of milieux. To be able to do that is to possess the sociological imagination. (Wright Mills, 1959, pp. 10–11.)

This difference between the two disciplines is reinforced by their attitude to time and the processes of change. Economics has usually begun from a consideration of static, equilibrium conditions to which time has later, and often only with great difficulty, been added (Lipsey, 1963, pp. 153–4; Richardson, 1971, pp. 13–14). By contrast, the sociologist almost inevitably finds himself closely concerned with social change and conceptions of historical development (Wright Mills, 1959, p. 6).

Economic analysis of the housing market

It is undoubtedly true that economists have found the city in general, and the urban housing market in particular, a peculiarly intractable problem, 'The economist armed with his kit of familiar tools is likely to be confused and bewildered when asked to dissect the city' (Richardson, 1971, p. 14). The reasons for this are manifold: they relate to the overall complexity of the city itself, the significance of temporal and spatial dimensions in urban behaviour, the great importance of external economies generated by the close inter-relationships of factors in the urban setting, the numerous opportunities for monopolies and oligopolies to develop, and especially the constraints and imperfections in urban markets of all types introduced by the spread of controls and intervention by public authorities.

It has already been suggested that the last of these points has been one of the major features that have distinguished the evolution of housing in Britain during this century. This has been seen both through the introduction of such controls as building bye-laws, town planning legislation or fixing the levels of rents which private landlords may charge, and through the direct intervention of local authorities in the housebuilding process. This has meant not only an increasingly 'imperfect' market in housing, but the creation of a situation where there exist a number of sectors in that market, whose operations are only poorly linked. One recent study of the relationship of the future levels of demand for housing and rises in real incomes stated,

A significant but seldom commented on fact is that unless one goes back to before 1915, only for a few years in the 1930s, and since around 1960 have more than one half of all British households lived in that part of the housing stock where the relationship between housing expenditure and income are even in principle market determined. (Byatt, Holmans, & Laidler, 1972, p. 3.)

and as they go on to illustrate, though there are theoretical structures to explain existing relationships in the housing market, it is virtually impossible to construct a model which will stand up to the strains of practical use. They summarise the situation by saying that, 'the British housing system is very different from such a hypothetical situation and likely to remain so' (Byatt *et al.*, 1972, p. 4). It is therefore not surprising that Cullingworth should have been able to comment on the lack of 'straightforward housing market analyses' in this country (Cullingworth & Orr, 1969, p. 186).

That situation may be contrasted with the United States, where though the housing market certainly has many complexities, the involvement of public authorities has neither been so widespread nor of importance over as long a period. There, the examination of housing markets and housing policy through economic analysis is an extremely well-developed subject (Grigsby, 1963; Muth, 1969; Perloff, 1973). Muth has stated the position of most American economists interested in the housing market: 'long-run comparative static analysis is a highly fruitful source of propositions which stand up quite well to empirical testing' (Muth, 1969, p. 95), and he explicitly contrasts this view with the contrary idea put forward by the British economist, Turvey:

It is impossible to present a comparative statics analysis which will explain the layout of towns and the pattern of buildings; the determining background conditions are insufficiently stationary in relation to the durability of buildings. In other words, each town must be examined separately and historically. The features of London, for example, can be fully understood only by investigating its past; it is as it is because it was as it was. (Turvey, 1957, pp. 47–8.)

In the past fifteen years, there has been a certain change of attitude amongst British economists and Kirwan and Martin have stated the basis for this new interest: 'if there is an active market in housing; and if the market works, it is worth attempting to understand it' (Kirwan & Martin, 1970, p. 11). Economic models of the housing market are now being developed in this country to analyse a number of particular elements and situations, as for example in the construction of a consumer surplus model to evaluate the costs and benefits of environmental improvements (Whitbread & Bird, 1973); in assessing the merits of rebuilding or renovation in older housing areas (Sigsworth & Wilkinson, 1967; Needleman, 1968; 1969); or in predicting the future relationship of housing demand and rises in real income (Byatt, Holmans & Laidler, 1972).

13

At the more general level of analysis, urban economists have concentrated on the construction of models of the spatial distribution of land values. The history of such exercises is a long one, dating back to Ricardo and Von Thunen in the early nineteenth century.[9] Modern work has concentrated on the development of an equilibrium land price surface from the aggregation of individuals' bid-rents for various parcels of land (Alonso, 1964),[10] where the key factor determining such individual bid-rents is often assumed to be the varying value placed on accessibility by different urban functions and different individuals.[11]

Much of this work bears a strong relationship in its aims (if not always in its methods) to the parallel development of spatial interaction models in geography and planning, and hopes have been expressed that a fusion of such economic and spatial models of the housing market may be possible in the future (Senior, 1973, pp. 193–4).[12] But if there are limitations in the construction of equilibrium models, those difficulties involved in the incorporation of dynamic elements into economic models of the housing market are far greater.

Given the economists' usual concern with the equilibrium of supply and demand, it is not surprising that the central process at work in the housing market over a long period of time is considered by them to be the way in which this equilibrium is maintained. It is an obvious difficulty, which has suggested itself already, that although changes in the social, economic and political climate must encourage, indeed require, parallel changes in individuals' attitudes to residential locations, the stock of dwellings itself alters only very slowly; more slowly in fact than the demographic structure of even the most stable area or city. It is therefore to be assumed that most properties are likely to experience relative changes in their values and standards during their lifetimes.

Even though the annual proportion of houses added or replaced in the total stock is small, the average life expectancy of property is so great as to lead to most dwellings experiencing a relative depreciation in value as they age because they become increasingly obsolete in the face of newer properties. Such obsolescence may take many forms. Most commonly, it derives from the improvement in facilities which rises in the standard of living encourage in newer houses, but obsolescence may also stem from fashion, as in the twentieth century desire for a garden, or most significantly from the point of view of this study, from a deterioration in the quality and condition of neighbouring properties.

In the most simple assumption, the average rate of depreciation in the relative value of a property should be related to its life-span. It has already been indicated that the average life of residential property in Britain, today, is about a hundred years.[13] Cairncross thought that most properties would suffer no depreciation in the first twenty years of their lives and would then depreciate at a rate of $1\frac{1}{4}$ per cent per annum till they reached 'site value' after a century (Cairncross, 1953). Jones and Clark using empirical evidence from the Institute of Valuers suggested an initial period of thirty years for nil depreciation, followed by an average rate of decline of $1\frac{1}{3}$ per cent per annum to reach 'site value' after 105 years (Jones & Clark, 1971).[14]

The process sounds delightfully simple, but in reality, the history of individual properties is far more complicated. Whilst it is indisputable that most property does depreciate in value as it ages (like most other goods), the speed with which this may happen will vary even for dwellings of an identical age and type. In particular, the character of maintenance of an individual dwelling will be of importance. 'A house does not collapse after one hundred years but rather its life can be lengthened or shortened substantially according to the treatment it receives' (Duncan, 1971, p. xi). It is even possible for the process to be halted or reversed. The phenomenon of 'gentrification' in many areas of Inner London has become very well known in the last few years, and is an excellent example of the complex manner in which changes in the framework of the housing market may be initiated and then worked through with unforeseen consequences. Here, fundamental changes in the social structure of many areas have been accomplished through an interaction of residential preferences and housing policy taking place in a specific locational context determined by the historical pattern of urban growth.[15]

This example is significant because it illustrates the connection between changes in the value of individual properties and the social structure and geography of urban areas. The whole process whereby properties slowly drift down the market, attracting progressively lower relative rentals, is commonly called 'filtering', and although some economists consider it important to restrict the use of that term to cover only such value changes (e.g. Grigsby, 1963), 'filtering' is now often used to encompass both economic and social changes in the character of properties and housing areas (Watson, 1971).

It is in this area that the need to develop a link with parallel work in sociology, and between the contemporary and historical considerations of such problems, is greatest. Having sketched in (if only in the briefest fashion) the economists' considerations of these problems, it is now appropriate to extend the argument by looking at the way in which sociologists have examined them.

Sociological analysis of the housing market

Empirically, the sociological significance of housing is well-established. Chapman (1955) undertook an extensive survey which investigated both the structure and material contents of a large number of dwellings and compared these with the character of the families living in them. He found a strong, positive correlation between the quality of the home and the social status of its inhabitants. That the home where most people spend the greatest proportion of their lives should be so closely related to such basic, social facts as income and status is hardly surprising. The enormous literature on social segregation that has been produced in the United States, and to a lesser extent in Britain, assumes the strength of such a relationship in individuals' residential location decisions (Duncan, Cuzzart and Duncan, 1961). Similarly, the extensive use of such property indicators such as rating valuations as surrogates for social ranking has been shown to have been very important in recent studies of British, urban, social geography.

It is rather curious that, after it has been suggested that sociology is a discipline which concentrates on the larger scale and interprets individual actions in the light of wider perspectives, so much of the sociological research into the housing market in Britain should have been at the micro-level. It has been typical of British studies that urban sociology should have been a peripheral part of the discipline, somewhat divorced from the main stream of sociological theory (Glass, 1955).

The two lines of research that have formed the bulk of 'sociological' studies of housing are investigations of social welfare problems at the local level and 'community' studies. The former have ranged from studies of the effects of relocating families on new housing estates after the redevelopment of slum properties (MHLG, 1970), of the social problems associated with New Towns and town development schemes (CHAC, 1967), of living in high-rise flats (Jephcott, 1972), or of the detailed relationship between estate design and the residential satisfaction of its inhabitants (DoE, 1972), to a number of studies concerned with social interaction, the relationships of neighbours and so forth (Bracey, 1964; Toomey, 1970).

The latter have had strong links with the other main line of research, that of community studies. The history of the community in urban sociology is a long and complex one, closely tied in with the theory of human ecology and having strong links with town planning through the idea of the 'neighbourhood' planning unit.[16] There have been a great number of such studies in Britain, both in rural and urban settings (Frankenberg, 1966; Kuper, 1953; Mitchell, 1954; Bott, 1957; Willmott & Young, 1957; 1960; Collison, 1959; 1963; Stacey, 1960). Though the workings of the housing market have not been a central concern of most of these studies, they have been important in acknowledging the significance of housing conditions in urban society (e.g Morris & Mogey, 1965).

Many of the 'community studies' have been concerned with social change, especially as it has resulted from the breaking up of such communities through redevelopment of older housing areas (e.g. Willmott & Young, 1957), but in a broader sense they have assumed a relatively static picture of urban social structure based on the existence of the sort of pattern of social geography represented by the idea of 'urban villages' (Gans, 1958), and studies like Willmott and Young's in Bethnal Green have been influential in presenting a picture of changes in the housing market being destructive of social cohesion (Willmott & Young, 1957, pp. 186–99).

Recently, however, there have been attempts to incorporate more general premises of sociological theory in a model of the housing market. From the point of view of the influence it has had, the most significant work has been the model of housing classes developed by Rex from a study of the situation of Commonwealth immigrants in the Sparkbrook district of Birmingham (Rex & Moore, 1967).

The Rex model begins with the assumption that housing is a scarce resource and will therefore be subject to processes of competition. It suggests that such competition will take three forms – for sites, for the use of existing buildings, and for welfare payments (as for instance in the case of subsidised council housing). These processes combine to form a special case of the general theory of class developed by Max Weber (Weber, 1948, pp. 181–94). Thus,

Max Weber . . . relativised Marx's theory of the nature of social classes by suggesting that any market situation, and not only the labour market, led to the emergence of groups with a common market position and common market interests which could be called 'classes'. We need only qualify this slightly to include groups differentially placed with regard to the system of bureaucratic allocation to arrive at a notion of 'housing classes' which is extremely useful in analysing urban structure and processes. (Rex & Moore, 1967, p. 164.)

Housing is thereby seen as a case in which a 'market' situation implying differential access to scarce resources creates a class system which will presumably reflect the values of, and distribution of power in, the urban social structure.

The Rex model has been acquiring some popularity both academically and more surprisingly perhaps as a practical tool in the administration of urban housing markets. Pahl and his students at the University of Kent took up 'housing classes' as a methodology capable of considering the allocative as well as the physical aspects of public planning policy (Pahl, 1968; 1970; Barbolet, 1969). MacMurray and Shoults (1973) have discussed the potential of a very similar framework for the analysis of the current housing situation in an inner London borough. Their reasons for adopting this approach were essentially practical. They found that,

in practice, current approaches to understanding housing matters have substantial inadequacies . . . Briefly, current approaches do not enable or encourage the analysts to develop a comprehensive view of the housing market under study, particularly within the operational constraints of local authority practice. (MacMurray & Shoults, 1973, p. 61.)

They continue by citing the lack of a consistent and usable theory of the housing market which allows consideration of all its aspects and they stress the practical need for a broad overview of all factors – physical, social and economic – when formulating housing policy. In particular, they point out that an approach which concentrates upon the identification of actors and agencies in the market process bears most resemblance to the political context in which housing policy decisions are taken.

Such a practical development of the Rex model illustrates its robustness for a more general consideration of the evolution of the spatial structure of the

17

city. It is true that the housing class model was not designed specifically to meet either spatial or temporal criteria, but Rex avowedly acknowledged the work of Burgess and other ecologists as of great empirical significance and stated as one of his initial purposes the placing of such ecological studies in a more rigorous theoretical framework. The associations of the housing-class model with the ideas put forward by the ecologists is obvious and it is equally clear that the forms of competition discussed by Rex, and especially those for sites and the use of existing buildings will have great spatial significance.

Temporally, it is paramount to recognise that change is extremely important in Rex's wider view of social conflict and he has gone so far as to suggest that his theory of conflict is a theory of social change (Rex, 1961, pp. 130–4). It should also perhaps be pointed out that others have stressed the role of social conflict in the spatial structure of the city, both empirically (Boal, 1969; 1972) and conceptually (Simmie, 1974).

The main limitation of the Rex model is its weak links with the obviously parallel economic processes indicated in the previous section. It has been criticised, in particular, because, 'any theory of housing as an allocative structure must explicitly incorporate the economic processes of the housing market' (Stewart, 1973, p. 209). From the viewpoint of this study, interested as it is in the interaction of economic and social factors and their translation through the housing market into the spatial structure of the city, this link is a key one. Though one might not so much be concerned with the manner in which conflicts in the social structure as a whole are resolved or ameliorated, it is extremely important to identify how any conflict-orientated pattern of social change may interact with the processes of economic change that have been identified.

The spatial evolution of the housing market

Harvey has suggested one way these links might be strengthened which does illustrate a methodology by which the interaction of economic and social changes might be examined in the long-term.

Harvey starts by emphasising the extraordinary practical difficulties of achieving any form of spatial equilibrium in the city. Though there must be processes (like those implicit in the economic evolution of the housing market) which will be moving towards an equilibrium at any one point in time, the total combination of processes and conditions that make up the spatial form of the city can rarely, if ever, be in harmony, and in any holistic sense, the concept of an equilibrium in so complex and dynamic a situation as the city is a nebulous one. Indeed, so complex are the forces at work in the housing market itself that it is equally difficult to visualise the practical existence of any equilibrium in this situation. Harvey characterises an urban system as being one 'in a permanent state of differential disequilibrium' (Harvey, 1972, p. 271), and this can be just as well applied to the urban sub-system of the

18

housing market, where it has been described how the interaction of physical and social variables with long-term economic processes provides a continually changing circumstance.

It follows that if one may identify the urban system as one in which no general equilibrium is present, but only a series of partial states undergoing continuous revision because of changes in other variables, then, 'the main question is the speed with which the different parts of an urban system can adjust to the changes occurring within it' (Harvey, 1972, p. 271). Obviously, scale is a key issue here (as it has been throughout the earlier parts of this discussion). Scale will be important spatially and temporally, but perhaps most significantly, there will have to be some reconciliation between the individual and the total structure of the city.

The factor which Harvey emphasises in considering this problem is one which has already been mentioned in connection with the difficulties which economists face in analysing the city – that of external economies. External economies,

may be said to arise when relevant effects on production and welfare go wholly or partially unpriced. Being outside the price system such external effects are sometimes looked upon as by-products, wanted or unwanted, of other people's activities that immediately or indirectly affect the welfare of individuals. (Mishan, 1969, p. 164.)

For the spatial structure of the city, externalities are of enormous importance.[17] Location in an urban area is almost entirely relative. There may be absolute environmental criteria related to some fact of the natural environment (costs of building, for example, may vary with detailed site conditions, e.g. altitude, drainage and so forth) but, for the most part, where one lives is of significance to one's everyday activities – travelling to work, to shop, to see one's friends and relatives. Therefore, since the city has already been characterised as a system in a state of continual change, it must follow that the costs and benefits of every location are also continually changing.

Indeed, one can go further, for Harvey follows a number of other people in assuming that a growth in externalities is the natural concomitant of increasing material wealth and thereby very strongly associated with the expansion of the urban system (Lampard, 1963, pp. 225–47). Not only will the city-dweller benefit from a much wider range and choice of facilities than someone living in a village, but also his total equation of locational costs and benefits is likely to be composed of far more elements, and thereby be much more difficult for the individual to assess or solve. Under such conditions of increasing scale, the greater number of actors in the urban system is almost inevitably going to be associated with a more dynamic situation, since even a transitory equilibrium is unlikely in so complex a situation.

A simplistic approach tends to ignore the effects of such relationships (or to assume that their net effects are nil), but in practice the efforts of residents to influence, retard or divert the pattern of change in externalities will be an important factor in the operation of the city. But the power of the individual will usually be far less than that of a group, and it is through the combination by residents to manipulate locational decisions that external economies are likely to exercise their greatest influence on the development of the city. Indeed, Harvey argues that the housing market,

must contain group action if it is to function coherently. This explains in turn why the housing market is so peculiarly susceptible to economic and political pressures since it is only by organising and applying these pressures that individuals can defend or enhance the value of their property rights relative to those of other individuals. (Harvey, 1973, p. 68.)

Obviously, such group action can take place on a number of levels. It can take the form of a ruling class or power group exercising political strength, or it can be merely the formation of a local pressure group, perhaps specifically intended to fight a particular threat, like a proposal for a new road scheme.[18]

Harvey's suggestions as to the role of group action are not identical to Rex's model of housing classes (they are more specifically locational in context), but there are obvious affinities, and it is easy to envisage circumstances where they coincide. For example, to return to the specific instance of the need to combat some immediate environmental threat to an area, such as a new road, it may well be that the response of individual householders will vary according to their housing class. A classic example has been the differing reactions of tenants and owner-occupiers to proposals to redevelop areas of sub-standard housing when rights to compensation and rehousing have varied between the two groups.

It is an attractive idea that the manipulation of external economies is one of the most important mechanisms which allows the social structure to influence the pattern of economic change within the housing market, and it does present at least some theoretical resolution of the problems posed by the various approaches already discussed. But there are at least two difficulties that suggest themselves. The first is the continuing one that some sort of operational framework for the long-term study of these processes is required, and it is by no means obvious that external economies are any easier to measure than any of the other concepts so far discussed.

A solution perhaps lies in the second unresolved problem. This is that the manipulation of externalities is insufficient to explain the pattern of change in the spatial structure of the housing market. If one accepts that that market is in a constant state of flux, and that consequently individuals' locational costs and benefits are under continual 'threat', there are three logical alternatives that they may adopt. The first has already been discussed: they may try (probably in a group) to influence those changes to their advantage. The second is that they will accept the changes even if they result in some loss. The third alternative is that although they may feel unable or unwilling to undertake to influence the changing pattern of externalities, they will seek by moving to restore themselves to a location where the net equation of costs and benefits is broadly similar. That movement or the influencing of externalities

20

are alternatives may be easily demonstrated by reference back to the economists' simple assumptions about the existence of an equilibrium in the housing market. Given growth in the city and a continuing pattern of change in locational values, it is obvious that to maintain such an equilibrium without equally frequent changes in the circumstances of consumers (i.e. so that the total equation of demand in the city will always match the equation of supply), all householders would have to be continually changing their addresses. That they do not do so will in part be a reflection of the way in which householders trade-off losses in locational value against the inconvenience of moving, and in part the result of success in manipulating the pattern of changes in externalities so as to minimise the need for such developments.

Residential mobility can therefore be identified as a process of paramount importance in the spatial evolution of the city, and through a long-term examination of mobility, considerable light may be thrown on the manner in which the various processes and mechanisms mentioned here actually interact.

Residential mobility

It is important to stress initially that residential mobility is itself a complex phenomenon, and the way in which it will be interpreted here will be only one aspect of its importance. For this reason, it is necessary to take some space to provide a context for the process and those ways in which it has been studied.

Table 2.2 illustrates current rates of mobility at various levels in Britain as indicated by the 1966 Census.[19]

TABLE 2.2　*Rates of mobility: 1966 Census*

	Percentage of the population	
Area	Moved in last year	Moved in last 5 years
England & Wales	10.0	33.1
East Midlands Region	9.0	30.5
Leicestershire	8.8	27.3
Leicester CB	9.9	28.4

But such averages conceal considerable variations. The General Household Survey identified a small proportion of people (about 1 per cent of their sample) who had moved as frequently as five times in the previous five years, and a very significant proportion (about 40 per cent) who had not changed their address for over a decade.

As a preliminary, it is necessary to establish suitable definitions. Residential mobility in the context with which one is here concerned – the evolution of the city's spatial structure – is part of a much wider pattern of behaviour. Firstly, there are those actions involved in entering and leaving the housing market. Households are being formed and dissolved all the time, and this is a very important element in the total changes which are going in on the market. 'Natural' demographic events such as marriage and death are major aspects of this, but they must accurately be seen as only part of the manner in which households are created and extinguished.[20] Recent surveys have suggested that about 30 per cent of moves are the result of the creation of new households, and that two-thirds of these come about through marriage (General Household Survey, 1973, p. 165). Similarly, at the other end of the demographic spectrum, not only will there be deaths, but also such events as retirement to old peoples' homes, going back to live with children and so forth.[21]

The movement of established households may be conventionally split into inter- and intra-urban categories. This is not an entirely arbitrary division. Simmons (1968) has suggested that an intra-urban move is one made with a radius sufficiently small not to require a change in one's workplace. Today, with huge metropolitan regions, and cities like London having a commuting field covering up to thousands of square miles, a move of more than fifty miles may be intra-urban by this definition.

In analysing migration, Germani (1964) has distinguished three scales of work. These are:

(1) the *objective*, which deals with the basic characteristics of the household which moves, its origin and destination and its age, social class and so forth;

(2) the *normative*, which considers the direct reasons cited for moving – to change one's job, to get a better home, etc.; and,

(3) the *psycho-social*, which investigates what social psychologists consider the underlying motivations for moving. These may relate to the individual personality of the mover and should throw light on the problem as to why with two persons in apparently identical circumstances, one will move and one will stay (Taylor, 1969).

All three levels of analysis have been applied to intra-urban movement.

At the objective level, the character of movers has proved easier to account for than the pattern of their moves. Rossi's study established a series of principles relating to the probabilities of moving which have proved very useful in the analysis of intra-urban mobility (Rossi, 1955). Such studies as Moore's which examine rates of mobility in small urban areas have illustrated the close parallels between the associations of movement in the intra-urban case with that in society as a whole (Moore, 1966).

There has been far less success in deriving such general principles which can be applied to the pattern of movement in the city. Classical explanations of migration, dating back to the work of Ravenstein in the late nineteenth century (Ravenstein, 1889), related to labour mobility and linked the strength of flows between two points to their relative attractiveness in terms of employment opportunities constrained by the intervening distance. Such work has been an important initial stimulus to the whole family of 'gravity' formulations that has played so great a part in the recent development of urban models and remains the foundation of research into movement at the inter-regional scale where considerations of the job market are paramount in determining the strength and direction of flows (Lind, 1969; Masser, 1970; Fielding, 1969).

But moves related to changing one's job and involving considerable distances are only a minority. In large urban areas, they probably do not exceed 20–25 per cent of all moves (e.g. West Midlands Housing Survey, 1972, p. 52), though they may form a higher proportion of moves in rural areas with more scattered and smaller units of settlement. Consequently, such classical forms of analysis have not proved so successful as a basis for evaluating the pattern of movement in cities. Limited size means that although distance may still be a factor in movement, it will almost always be outweighed as a means of explanation by the many other considerations that govern the choice of where to live. Urban structure and the pattern of the housing market create a set of constraints which severely complicate movement.

One method adopted in some American studies to overcome these problems has been to try to adapt the methodology of labour-mobility models through the concept of the mean information field (Marble & Nyusten, 1963). The mean information field is the overall assessment of the conception of distance which the inhabitants of an area are thought to possess and which thereby influences all their regular forms of spatial behaviour. The pattern of migration is only one symptom of this generalised distance function. The mean information field was originally derived from the calibration of a model based on an enormous amount of empirical data, and later studies have attempted to substitute this by some other surrogate such as marriage distances or shopping patterns (Morrill & Pitts, 1967; Clark, 1969; 1970). Links between migration and other forms of behaviour in the city have also figured in studies of the role which it plays in the spatial growth of the city (Dewey, 1948; Ross, 1961–2; Schnore, 1964; Goldstein & Mayer, 1965–6).

Building a comprehensive picture of the relationship of migration to urban structure has been much more difficult. Boyce (1969) recognised this difficulty, but Adams' studies of the migration patterns in Mid-Western American cities and their interpretation in the light of the classical ecological models are some of the few substantive examinations of the problem (Adams, 1969; 1970). In particular, Adams showed the relevance of the age of building and the concentric distribution of house types in directing flows.

The importance of housing is comforting from the viewpoint of this study and leads on to the consideration of the normative level of analysis – the reasons that people give for moving. Most modern surveys have contained a section which discusses the reasons given either for recent moves or for a declared intention to move. For example, the General Household Survey used five main groupings in its assessment of movement: (a) housing, (b) environmental, (c) work, (d) personal and (e) other. Work moves accounted for 17 per cent of the total, which ties in with the significance of longer-distance flows. The most important single category was personal, under which heading were all those moves associated with the creation or dissolution of households, and this too is supported by the evidence of the importance of this factor in the total pattern of residential mobility.

From the special standpoint of intra-urban migration and its relationship to the housing market and the evolution of the city's spatial structure, the most interesting categories were those moves cited as being concerned with housing and the environment. The General Household Survey suggested that housing reasons were the major factor in over 25 per cent of all moves, and environmental reasons in about 10 per cent, but unfortunately it is extremely difficult to provide an adequate definition of either category. Any move involves changing one's house, and even if all the characteristics of the two dwellings are very similar, the whole of the preceding argument concerning the significance of externalities depends on the impossibility of totally replicating the previous equation of locational costs in one's new residence. Therefore, all moves are in a very fundamental sense, housing moves. Similarly, in describing a move as caused by a wish to improve one's residential environment, it is usual that a very complex set of causes and motives is subsumed within the most simple statement that one wishes to seek a better area in which to live, which may be a necessary, but is hardly a sufficient condition to describe the reasons for moving.

It is therefore not surprising that in the last few years, attention should have turned to focus on the third of the three levels of analysis that Germani suggested – the psycho-social. In the urban context, this has meant a much greater concern (paralleled in many other studies of spatial processes in the city) to identify the conditions that determine the pattern of decision-making in migrant households. Wolpert's model of environmental stress as a trigger to movement argues that it is the combination of individual circumstances with the changing character of the environment that induces the decision to move, and this study has been followed by a number of others that concentrate on the conditions that influence movement in urban areas (Wolpert, 1966; Scott, 1969; Moore, 1969; Brown & Moore, 1970; Brown & Longbrake, 1970; Brown & Holmes, 1971).

This sort of work will continue, and may be very useful in moving towards a general model of spatial decision-making in urban areas, but it is at present focused on a level which is not especially useful to the scale with which this study is concerned. What is significant, however, is the manner in which it has stressed that mobility is a consequence of a conflict, resolution, or marriage of personal and environmental causes. It has been suggested that all residents need to trade-off the possible diseconomies of remaining where they are at present living against the costs of moving when faced by alterations in the structure of external economies, and it is extremely interesting that most of the studies of mobility that have been carried out in urban areas have tended to operate with the assumption that there are in fact two sets of controls, one personal, the other environmental, that will influence the pattern of movement. Ross, for instance, thought that short-distance moves within the city

24

were personally controlled, whilst the longer-distance movements were environmentally determined (Ross, 1961–2).

In practice, it has always been far easier to specify the structure of influences on the personal side than the environmental. From Rossi's study onwards, there has been an emphasis on a relationship of mobility to demographic and social factors which is most fully explicated in the life-cycle/career-mobility model (Leslie & Richardson, 1961). This derives from two very well recognised associations of rates of mobility in contemporary, Western, industrial societies.

Firstly, it has been observed that the probabilities of moving house decline with age (Goldschreider, 1966). The General Household Survey, for example, found that only 15 per cent of householders aged over sixty-five had moved at least once in the previous five years, compared with 80 per cent for householders under thirty years of age. This correlation has led on to the construction of a 'life-cycle' whereby peoples' ages may be associated with their accommodation needs:

(1) For the first twenty years or so, they live with their parents, and therefore do not form independent households.

(2) Then a growing proportion of them spend a brief period on their own or with friends, after leaving home to study or to find work. The first year or two of marriage, when wives generally remain at work, may be regarded as a continuation of this phase; the household is small and mobile, and out all day.

(3) As soon as its first baby is born, the household's needs change again and become, during this expanding phase, increasingly extensive and demanding.

(4) In time, all or most of their children leave home, and for those who do not have elderly relatives living with them, there follows a fourth phase. The household is again small, and less dependent on its neighbours and the services afforded by the surrounding neighbourhood, but a home has been established and filled with possessions, roots have been put down, and people are less likely to move than in earlier years.

(5) Finally, in old age, households shrink still further; they become even less mobile, and their comfort and peace of mind depend increasingly upon security of tenure, upon the design and equipment of the home, the services available in the district and the support of nearby friends and relatives.[22]

The career-mobility section of the model derives from the modifications to the patterns of mobility suggested by the life-cycle produced by a well-observed, strong, positive correlation between movement and social class. Musgrove (1963) has gone so far as to coin the phrase the 'migratory elite' to describe the growing propensity of professional men and executive staff to have to conduct their careers within the setting of national organisations, both in private companies and the public service, where promotion prospects may depend upon a willingness to move. The General Household Survey gave very clear evidence of this trend. Whereas 51 per cent of professional men had moved at least once in the previous five years and 20 per cent at least twice, by contrast, only 30 per cent of semi-skilled and unskilled workers had moved once, and less than 10 per cent twice (General Household Survey, 1973, p. 159).

Leslie & Richardson (1961) combined the life-cycle and career-mobility elements into a 'paradigm' for residential mobility which stated that those who would be most likely to have an intention of moving would be persons in the expanding stages of the life-cycle (i.e. stages 2 or 3) and have upward mobility potential (i.e. they would be in careers which probably require qualifications, carry a higher social status and might be expected to require some degree of geographic mobility): those least likely to have an intention to move would be in the non-expanding stage of the life-cycle (i.e. especially stages 4 or 5) and have little upward mobility potential (i.e. they would be in unskilled or semi-skilled jobs where there were little prospects of promotion). The model has been tested and used in a great many studies and its generalisations have stood up well to empirical testing (Jansen, 1969), but it is extremely important to realise that the life-cycle/career-mobility model is only a very general explanation for mobility in urban areas, which is better treated as a set of associations which underlie the various detailed reasons as to why households move.

This point is especially reinforced by reference to the other side of the equation of mobility, the way in which environmental changes generate movement. These are often cited as being very important factors in intra-urban mobility and migration, but as has already been suggested, it is extremely difficult to present any form of generalisation as to their operations which parallels the life-cycle/career-mobility model's role in describing movement's correlation with personal factors.

That there is some association between movement and changes in the environment of urban areas lies at the heart of much of what has been discussed in this introductory section. The whole argument about the role of external economies and the relationship of these to the progress of changes in the relative value of properties in the housing stock implied by the 'filtering' model assumes such a relationship. But it is very difficult to isolate moves which are specific examples of such responses. The best known of such attempts is the ecologists' concern with the concept of 'invasion–succession'. This was one of the ideas borrowed from plant biology, whereby successive species occupy a particular habitat in a certain sequence, and was used by the ecologists to explain the manner in which growth in the urban area would be translated into changes in the social structure of local parts of the city. It assumed that as residential districts filtered down the housing market, they would be occupied by a sequence of social or ethnic groups, and that the passage of the area's population from domination by one group to domination by another would be accomplished by a repeating pattern of changes. In particular, it assumed that residents would possess a high degree of awareness of the character and quality of the districts in which they lived, and that consequently, any initial 'colonisation' of the area by members of another, lower, social group would be followed by a rapid and conscious exodus on the part of the original inhabitants (Burgess, 1928).

The 'invasion–succession' model of social change in small areas has been enormously influential in studies in the United States, especially where the sequences observed have not been social so much as ethnic. The study of racial groups' movement through the city, of degrees of segregation and of such special factors as 'tipping ratios' (i.e. the point at which the conscious movement out of an area's previous inhabitants is triggered off by the colonisation of some new group) have been enormously important in American cities (Lieberson, 1963). In practice, the direct evidence for such processes in the United States has recently been called into question (e.g. Molotch, 1969), and it has been pointed out that there are a number of difficulties inherent in the operation of such a model.[23]

Of course, such concerns with the distribution of ethnic groups have traditionally been of much less significance in British cities until very recently (e.g. P. N. Jones, 1967; 1970; Lee, 1973; Rowley & Tipple, 1974; Kearsley & Srivastava, 1974), and it has already been demonstrated that surveys of household mobility do not seem to place great importance on such environmentally generated moves as factors in the total pattern of intra-urban migration. Indeed, it is not without significance that many recent studies in the United States have found much the same sort of proportion of moves in the city, where it is possible to ascribe directly motivation to environmental factors (e.g. Greenbie, 1969).

Such evidence suggests that any artificial division of movement into those which are environmentally significant and those that are not is likely to be quite worthless. Any understanding of the role which residential mobility plays as evidence of part of the interactions of physical and social changes in the geography of the city can only be arrived at by as complete a consideration of the process as is feasible. It would be foolish to try to seek to isolate only those moves which *a priori* one might consider as significant for such changes. Not only is residential mobility so multi-faceted a process that the substitution of any single motive is highly suspect, but it is clear that the total relationship of mobility to changes in the spatial structure of the city has to be considered in terms of the relationships between mobility and immobility, and that a decision not to move, or a set of actions designed to put off the need to move, are likely to be as important as movement itself.

The structure of the study

The study is based on an examination of the relationship of residential mobility, the housing market and the evolution of the spatial structure of one

English city, Leicester, since the Industrial Revolution, and is divided into three parts.

In the first, the situation in 1870 is examined in detail. 1870 is not an entirely arbitrary date: it does represent an important period in Leicester's history, when the city was in the midst of enormous social and economic changes, typified by the fact that it was during this time that Leicester's population was growing most quickly. It is also the period from which the oldest dwellings still occupied in the city today date. In this first part, the one year, 1870–1, is used to paint as complete a picture as possible of the state of the city's housing market and its relationship to contemporary patterns of residential mobility and intra-urban migration.

The second part of the study is a longitudinal approach to the relationship of residential mobility and urban growth. It is based on that set of city directories produced for Leicester between 1870 and 1940, and might be said to cover the 'dark age' of British research into urban areas. There is an enormous amount of work going on in current British cities, and there has also developed in recent years a growing interest in the structure of the Victorian city. The opportunities presented by the release of the Census Enumerators' original returns up to 1871 and the vast amounts of material now held in local authorities' record offices have stimulated a great interest in the period 1840 to 1870, but work in the period between then and 1945 is much, much rarer. For example, there has been considerably less research into the extremely important period between the two World Wars (e.g. Jackson, 1973).

The third section considers the modern period, looking at the development of the city between 1945 and the middle 1960s. It again concentrates on a single year, 1963–4, and attempts to relate the modern pattern of mobility and migration to the much altered post-Second World War housing market.

Each of the three substantive parts of the study takes a similar form, beginning with a brief contextual introduction to the economic and social condition of Leicester in each of the relevant periods. The first sections of each part are straightforward and follow fairly well-trodden paths of analysis of the city's social geography, with the proviso that they are especially slanted towards the structure and operation of the urban housing market. That market is considered through the identification of actors and actions at work in the market on both supply and demand sides, the legislative framework within which the market is operating, the significance of contemporary rates of building, and the role of such new development in meeting the demands of the market. The consideration of the housing market will particularly try to identify those factors which might act as barriers either to entry by new households or to movement within the city by existing.

Empirically, the most interesting section is that which deals with residential mobility and the pattern of intra-urban migration. As an initial proviso, it must be pointed out the manner in which these processes have been examined has been greatly determined by the paucity of usable data available which is relevant to mobility and migration in the city.[24] Consequently, it has been inevitable that the study be framed in terms of the simplest concepts and most elementary forms of analysis. In particular, three factors will be stressed in these processes.

The first of these will be simply the level of mobility. Mobility will be defined here in the widest sense so that all the reasons which can lead to changes in the occupation of dwellings will be included – be they the creation or dissolution of households or movement both within and into and out of the city. Mobility will not be treated as a constant throughout the city, but variations both between areas, housing types and differing groups of the population will also be considered.

Secondly, the connectivity of the urban system as revealed by the pattern of movement will be examined by looking at the degree of linkage between the various types and areas of housing in the city. Because the central concern here is the relationship to the housing market, this approach will be used in preference to the more familiar consideration of such geometric parameters as distance or direction.

Thirdly, wherever possible, the individual move will be considered in terms of what light it throws on the existence and persistence of such general models as the life-cycle/career-mobility paradigm and the ideas of invasion–succession. This does not imply a prejudgement that such ideas and relationships which may exist at present have been constant throughout the period under consideration, but they do form convenient pegs upon which to hang discussion of such phenomena.

From this examination of such elementary characteristics of the mobility and migration system of the city, it will be necessary to make the vital conceptual leap to infer the changing relationship between that system and parallel developments in the city's housing market and thereby to evaluate some of the key links in the association of urban social and spatial structures. The fourth section of each substantive part will attempt this and will hopefully contribute to a re-examination in the study's final chapter of some of the ideas sketched out in this introductory framework.

Leicester[25]

Leicester is an industrial city in the English Midlands, some 150 kilometres north-west of London. The City of Leicester was formerly a County Borough, but since 1 April 1974, it has been a District Authority within the 'new' county of Leicestershire. In 1971, the City had a population of over 284,000 according to the Census, but 'Greater Leicester' (i.e. the City plus its immediate surrounding suburbs) contained about 400,000 people living within ten miles of its centre. Leicester is also the centre of the county, which is inhabited by about three-quarters of a million people.

Leicester does have certain characteristics which make it especially suitable for this study. Firstly, it is unusual in English cities of this size (though not unique) in being exceptionally free-standing, 'the city and county of Leicester together constitute a classic case of the city region, a functional region composed of the dominant city performing central place functions for a wide hinterland, both urban and rural' (Hall, 1973, p. 560). There is now a considerable overlap between the commuting fields of Leicester and its industrial neighbours, especially with Derby and Nottingham in the north, and with Coventry in the south-west, trends that have been of greater and greater importance over the past twenty years (Osborne, 1954), but it is still reasonable to recognise a 'Leicester Region', centred on the city, which is

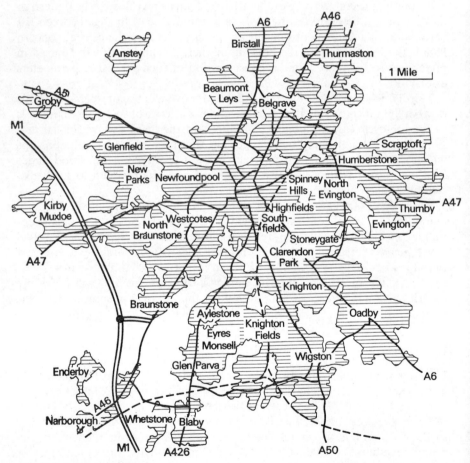

Fig. 2.1. Leicester, 1974

sufficiently distinctive to prove a great aid in examining the interactions between the housing market and local residential mobility links.[26]

Secondly, the continuing growth of the city so steadily over the past century has provided Leicester with a mix of housing that is reasonably representative of the main types of development that have occurred since the middle of the nineteenth century.[27] Prosperity and continuing building has meant that the city never accumulated the same legacy of large amounts of slum property that have occurred in some cities. Indeed, Leicester has been one of the best-housed of English cities (at least since the turn of the century) and this is in itself a good reason for choosing the city as a case study. The pathological elements of cities always attract attention; for the most part, Leicester represents the opposite side of the coin, and the dynamism of the city and the speed with which its physical structure has had to adapt to growth means that it is a peculiarly suitable example for tracing the relationships between physical fabric and social structure over a long period of time.

Thirdly, the city's industries have been predominantly light and 'clean' and the many small units of production that have been associated with its staple trades (such as hosiery or footwear) have meant that industry has never had a dominating influence over the city's spatial structure. Leicester has the advantage (from the viewpoint of this study) of being a city in which the locational patterns of housing have been particularly the result of forces and relationships within the housing market itself, and where the environment in the broadest sense has not preconditioned the social character of areas.[28]

Finally, Fig. 2.1 shows the city in 1974, marking the extent of the built-up area and the major districts within the city which will be mentioned in the text.

CHAPTER 3

HOUSING AND MIGRATION IN LEICESTER IN 1870

Choosing 1870 as the starting point for this study was dictated by a number of circumstances. It is a most suitable date from the point of view of available information and also the best starting point for comparison with modern conditions since much of the oldest inhabited housing in the city at the present time was built in this era. Most of all, 1870 lies in the middle of an extraordinarily important period in the development of the city. The most obvious elements of the changes were a phenomenal rate of growth in population and boom conditions in the local economy, but these were the 'tip of the iceberg'; underneath were parallel developments in its social structure, urban institutions and patterns of behaviour that caused Simmons to state, 'between 1850 and 1880, modern Leicester was created' (Simmons, 1965–6, p. 41).

Leicester is situated on a gravel shelf on the east side of the Soar Valley, at a point where there is easy access to a ford connecting with the ancient track of the Fosse Way, which runs north–south along the slopes of the western bank (Dury, 1963). As Ratae Coritanorum, Leicester was a Roman cantonal capital (Richmond, 1955; Peek & McWhirr, 1972) and the medieval town grew up on the same site. By Tudor times, Leicester had a population of about 600 households (Hoskins, 1957) and,

The opening of the seventeenth century found Leicester a small market town, albeit the biggest in the county, with a population of some 3,500. The town possessed its open fields, the inhabitants still followed largely agricultural pursuits and what industries were to be found in the borough were tied very closely to rural activities. (Martin & Newman, 1972, p. 280.)

A revolution in Leicester's role came with the introduction of the hosiery trade into the East Midlands, which converted the agricultural market town into a thriving centre of manufacturing. Undercutting and then destroying the power of the London guilds, by the beginning of the eighteenth century, Leicester was, with Nottingham, the administrative and commercial heart of the national hosiery trade.[1]

Though hosiery had its ups and downs like all other industries, eighteenth century Leicester was a prosperous community, and its thriving commerce was reflected in a considerable physical expansion in the latter half of that century.

Until after the middle of the nineteenth century, the hosiery industry continued to provide the economic foundation of this expansion. It lured migrants from the villages of the East

Midlands by the prospect of a trade that was relatively easy to enter and to learn, that required no capital outlay (as frames could be rented), and offered employment in conditions of relative independence – at least not subject to the discipline of a factory – not only to the framework-knitter himself but to his whole family, who could be occupied with the ancillary tasks of winding, seaming and footing. The system encouraged men to marry young and embark on families at an early age. (Evans, 1972, p. 289.)

But by the early years of the nineteenth century, the character of the hosiery industry had become anachronistic. It is clear that by 1815, Leicester was 'not as important as in former times' (Fowler's Directory, 1815, p. 1), and in the same year, a visitor was struck by the already unusual fact that there were 'no chimneys and no rumble of steam' (quoted by W. G. Jones, 1890). The city's decline was related to the old-fashioned manner by which the means of production in the industry were organised.

The East Midlands hosiery industry in the first part of the nineteenth century has been described as 'proto-industrial' (Head, 1961–2; Chapman, 1965). In very simple terms, there were three strata. At the top were the hosiers – the entrepreneurs who provided the capital which fuelled the system. Sometimes, these hosiers owned premises for working the yarn and employed a regular labour force, but more often they used middlemen, factors or agents to lease yarn, premises or frames to the bottom level in the industry, the individual framework-knitters. Whole families were used in the manufacturing process and often lived in dwellings which served as both home and workshop (Chapman, 1971). Leicester was the commercial centre of the industry, but manufacturing was carried on over a large area of the county, both northward along the Soar Valley and to the west and south in and around small towns like Hinckley (Smith, 1963). By 1790, it has been estimated that there were about 6000 persons engaged in the trade in Leicester itself, and the period between 1781 and 1811 has been described as the 'golden age' of the hosiery industry (Mounfield, 1972, p. 364).

After the end of the Napoleonic Wars, prosperity began to wane. Altogether, between 1750 and 1840, the number of knitting frames in Leicester had quadrupled (Nichol, 1924), and the consequent overcapacity in the industry brought under-employment and more extremely, unemployment. The city's labour market became hopelessly overloaded as the hosiers tried to keep down costs by importing cheap, unskilled hands, firstly from the surrounding rural hinterland, and later from as far afield as Ireland (Redford, 1926, p. 91). Wages slumped such that the aphorism 'As poor as a Leicester stockinger' became common (Evans, 1972, p. 289). The height of the crisis in the hosiery industry and in the city was reached in the 'Hungry Forties' when a national depression led to an economic and social calamity in Leicester as the bottom dropped out of the trade. In the winter of 1847–8, it was claimed that as many as a third of the population was forced to seek poor relief (Chamberlain, 1861). Little wonder that Chartism should have found the city such fertile ground (Patterson, 1954; Harrison, 1959).[2]

Leicester in the 1840s was something of a paradox. It was a city not suffering so much from the birthpangs of industrialisation, as from the death-throes of the old economic and social order. Many of the side-effects that had been associated with industrial growth elsewhere in Britain were plainly visible in the city. Between 1801 and 1861, Leicester's population had trebled (Smith, 1965, pp. 27–53; VCH, 1958), and though such a rate was considerably less than that for some of the northern industrial towns in the same period (Robson, 1973, pp. 38–9), it brought with it most of the same problems of overcrowding, poor sanitation and the general crisis in urban public health with which this period is so associated. Until the late 1850s, Leicester got most of the disadvantages of the Industrial Revolution without too many of the benefits in terms of economic growth and increased job opportunities.[3]

The situation began to improve in the 1850s, partly from the general up-swing of trade in the national economy, more especially from the very rapid development in the city of new industries. The depression of the 1840s brought the seeds of change to the hosiery industry itself. Many of the smaller hosiers went out of business, and those who survived took production firmly into their own hands. Steam-powered looms came to Leicester in 1845 at Thomas Collins' workshops in Bedford Street, and between 1845 and 1865, many more power-operated frames were introduced, culminating in the latter year with the opening of Corah's giant (for those days) St Margarets factory which symbolised the new footing upon which hosiery had been placed. Hand stockingers survived until the 1890s, but from the 1870s onwards, hosiery became an industry rather than a craft (W. G. Jones, 1890).

The most important of the new trades was the boot and shoe industry. The growth of the footwear trade in the city will always be associated with the name of Thomas Crick, whose development of new processes in the 1850s made Leicester into one of this industry's most important centres in the country (Mounfield, 1966; 1967). Prosperity began to pick up in the late 1850s (Annual Report of the Leicester Domestic Mission, 1859, p. 12), and in the next decade the city experienced boom conditions as one of the fastest growing large towns in England. Contemporary observers were well aware of the scale of the changes that were taking place. The Superintendent of the Leicester Registration District felt obliged to append a substantial note to explain the city's very rapid recent rate of growth when he reported to the 1871 Census,

The old manufactures have not increased, but have diminished in the last ten years. About the year 1861, the strike at Northampton caused the removal of a large portion of its shoe trade to Leicester, and the depression at Coventry a year or so later brought a large number of ribbon weavers from that city and its neighbourhood who were absorbed by the elastic web trade. Prior to 1861, the principal manufacture was hosiery. Since that time, new trades have been introduced. The shoe trade has brought sewing machine and nail makers, and others have arisen in connexion with the elastic web trade. There has been a large immigration from other places and it is still going on. (1871 Census, vol. II, p. 344.)

The 1871 Census found 95,364 people living within the boundaries of the city,[4] and the population of Leicester had increased by 40 per cent during the previous decade. There were just under 20,000 dwellings in the city, of which about a quarter had been built in the previous ten years, and the structure and organisation of the housing market that they were part of will now be investigated.

The Victorian housing market

Before attempting any discussion of the Victorian housing market, especially one which will of necessity be rather superficial, two preliminary points need to be made. The first is that to speak of a 'Victorian' housing market is almost as great an oversimplification as the 'Victorian' city. With the benefit of hindsight, one can of course recognise features common to the nineteenth century which were very different from those that exist today; but even the most basic changes (e.g. the whole question of the degree of involvement by public authorities in housing matters) should more accurately be interpreted as relative differences which themselves altered greatly during the whole course of the nineteenth century.

This temptation to 'concertina' one's appreciation of the housing market is intensified by the second preliminary point. This concerns the paucity of information available which was directly concerned with the structure and operation of the housing market. Chapman has pointed out that, 'the history of housing was only marginally covered in the familiar sources, and the only adequate source material was scattered through any quantity of estate papers, local deeds, maps, official reports, reformers' pamphlets, provincial newspapers, builders' accounts, fading photographs and many other sources' (Chapman, 1971, p. 9). Indeed, one might go further and suggest that much of this material, vital though it is, concentrates on property ownership rather than property occupation. In Leicester, like most nineteenth century cities, very few houses were owner-occupied, and the forms and negotiations which actually controlled the occupation of dwellings were often, probably usually, transitory, unrecorded and now only hinted at in available documentary material.

An inevitable consequence of this is that much more attention has been focused on the way in which housing formed part of the general process of urban growth in the nineteenth century (Dyos, 1961). This is not inappropriate. After all, the dominating characteristic of large urban areas in this period was their rapid growth. But it has meant that much more is known about how housing was initially provided and financed than about what happened to property after it had been built and was first occupied or the role of the housing market in relation to changes in the city's social and economic structure.

It is not surprising that much recent work which has looked more closely at the actual operations of the housing market should have concentrated on the great mass of literary and qualitative evidence and thereby should have emphasised the pathological symptoms of the Victorian housing market. The majority of social observers whose work is best remembered today, from Engels and Dickens in the 1840s through Mayhew and Octavia Hill to Booth and Rowntree at the end of the century, interpreted the condition of housing as evidence of a more general malaise in urban society. Obviously, such evidence has to be interpreted especially carefully in the light of the warning given at the beginning of this chapter and one must also be aware of the differences between the cities of Victorian England. To quote the most obvious example, there is a natural and growing interest in housing conditions in London (e.g. Wohl, 1971; 1973), but the situation in the capital seems likely to have been as much of a law unto itself then as it is today.[5]

Everywhere, though, the housing problem created by industrialisation was a gigantic one. The very rapid growth of large towns brought an enormous upsurge in demand if measured by the rising population, but most of this came from poor immigrants or their children who had taken up unskilled work earning minimal wages in industries where productivity was still low. The translation of such growth into an effective demand for housing was not easy and builders could afford to provide only the most rudimentary forms of shelter:

Suitable housing did not exist and the additional numbers were crammed into every nook and cranny from attic to cellar of old decaying property, or into cottages run up hastily in confined spaces with little or no access to light and air. (Kitson Clark, 1965, p. 79.)

Costs had to be kept to a minimum for customer and supplier.

Working class districts were built purely as a commercial undertaking which had to compete for capital with the most remunerative, alternative investments; they were supplying a demand at the level at which it was effective. (Ashworth, 1954, p. 20.)

The results were hardly surprising.

In cities like Leicester, which had expanded very quickly from a relatively small, pre-industrial base and in which a lot of the older, centrally situated dwellings would soon be pulled down to make way for expanding commercial and business functions, opportunities to meet the demand through the sub-division of properties were limited. Though there are no accurate figures for housebuilding in Leicester prior to 1849, evidence from the Census suggest that between 1800 and 1850, the total number of dwellings in the city quadrupled. To achieve minimum costs, the bulk of such property had to be as small as practical and as tightly packed together as the builder could manage. The first (if not the only) requirement of housing was that it be cheap, 'and pressure upon housing space was so great as to force up site values.' (Allan, 1965, p. 599). In the early days, 'urban densities increased at the expense of social segregation. Small cottages were squeezed between more substantial houses.' (Pahl, 1970, pp. 38–9.) This was especially well demonstrated in Leicester in the first half of the nineteenth century. In 1800, Leicester had had a reputation as an 'open' town, but the next forty years saw higher densities as continuous infilling built over existing yards and gardens with small groups of tiny cottages. The most extreme examples of this could be seen in the way in which the large yards behind the major coaching inns which fronted the main thoroughfares of the town were built over after the railways came and ruined their business.

By 1840, there was very little open space remaining within the confines of the historic town and building had already begun to spread outwards. Construction took the form of courts and yards appended to extensions of existing street lines fronted by short terraces. Courts were not as common a feature of the pattern of development in Leicester as say in Manchester or Liverpool, but even so, a survey by the first Borough Medical Officer of Health in 1848 found 347 courts in Leicester containing about 2000 dwellings, a sixth of the total in the city. Such courts contained houses 'upon an average with four rooms each, i.e. two bedrooms, a room for day occupation and a kitchen; the dimensions ranging from 12 feet to 14 feet by 8 feet to 10 feet, but there are no arrangements for ventilation.' (Ranger, 1849, p. 8).

Yet so far as densities were concerned, Leicester was considered a well-housed city by contemporary standards (Martin, 1845, p. 34). The city was fortunate not to suffer from the acute shortage of housing land that afflicted a place like Nottingham, which only enclosed its open fields comparatively late in the nineteenth century (Edwards, 1964). In Leicester, enclosure came early and ensured that there was adequate room for the outward expansion of the city (Hartopp, 1933; Evans, 1972).

Leicester's main problem in the first half of the nineteenth century was the public health one generated by an uncertain and contaminated water supply and inadequate drainage and sanitation. The city's site in the flood-plain of the River Soar brought obvious problems. As late as 1881, it could be said that, '"The floods are out!" has been a chronic cry from time immemorial.' (Read, 1881, p. 18). Frequent flooding combined with the hopelessly inadequate fall for drainage and sewage to provide appalling conditions for health and hygiene, 'A vast number of our community obtain their supply of water from wells in the vicinity of privy cess-pools, having no drainage into the public sewer' (Medical Officer of Health's Annual Report, 1859, p. 120).[7] Within the courts and yards of the poorest districts, such were the conditions of sanitation that they could be said to be, 'thoroughly innoculated with liquid refuse and effluvium from privy pits' (Ranger, 1849, p. 11). Such areas were obvious breeding grounds for disease. Fever and consumption were endemic, infant mortality rates were usually over 30 per cent and such districts suffered periodic outbreaks of cholera and smallpox right into the third quarter of the nineteenth century.

It was this association with disease that caused the 'urban problem' of the middle of the nineteenth century to emerge with such force in the 1840s when 'King Cholera' demonstrated that sickness could not be segregated, and social unrest and economic depression persuaded the influential segment of public opinion that a comprehensive approach to urban hygiene was required. Some of the quotations as to contemporary conditions in Leicester that have been given in this chapter come from the official inquiries of the 1840s that led to the passing of the 1848 Public Health Act.[8]

Ashworth has warned of the dangers of overestimating the significance of the 1848 Act (Ashworth, 1954, p. 58). It did enable the new industrial towns to take their affairs more into their own hands (Best, 1971, p. 39) by delegating to them powers to provide such public services as drainage and sewerage, water supply and street cleansing;[9] to act against conditions injurious to public health such as bad sanitation, 'noxious trades' and other nuisances; and to exercise some control over new development through statutory notification of new building.[10] The willingness of authorities to act within their powers varied greatly (Ashworth, 1954, pp. 47–77) and coercive powers even if they did act conscientiously were very limited.[11]

Leicester's general reputation in the field of sanitary reform, if not an outstanding record, seems to have stood as high as most cities, but progress was slow. In 1845, the City Corporation appointed an Improvement Committee to report on the possibilities of introducing a drainage and sewerage system and providing an unpolluted water supply. The Committee's report in 1846 set in motion a series of attempts to control the River Soar which were to continue until the 1880s and introduced the city's first sanitation system. A company to carry this out was floated in 1848 with the backing of the Corporation, but success was not immediately forthcoming. Though there were steady improvements in drainage and sewerage facilities, by the late 1860s only a third of all dwellings in the city were connected to the main drains, and complete coverage was not achieved till the mid-1890s (Windley, 1917).

The rather ineffectual building regulations of 1848 were strengthened in 1859 and further tightened by a Local Act of 1868 (Tansey, 1970). The gradually improving standards of new building had a number of connected effects. Firstly, they increased the cost of building houses. Better quality involved improved amenities and especially a more complex infrastructure – roads, paving, main drainage, sewerage and so forth. It therefore determined a major change in the morphology of development. The pattern of tightly packed courts, yards and short terraces gave way to the familiar grid of long terraced streets with separate back-yards. Falling densities meant a need for more land and thereby involved a more rapid peripheral expansion of the city. Also, the combination of these changes altered the relationship between new housing and the social structure.

If in the early part of the nineteenth century the very scale and character of the rapidly rising demand for accommodation in industrial cities determined that the quality of much of the new building would be very low, in Leicester between 1860 and 1875 the success of building regulations in raising the quality of new construction seems to have been a function of the city's growing prosperity and its consequent ability to generate far more households capable of purchasing a higher standard of shelter. Because Leicester was a city which grew more quickly in the second half of the nineteenth century than in the first, which was unusual, its period of greatest expansion coincided with these improvements. There seems little doubt that the economic expansion of the 1860s, with new jobs and greater prosperity, created a far larger number of families in the city who were able to afford the standard of property that the proponents of public health reform required. Leicester's great characteristic as a city in the last half of the nineteenth century was that its prosperity enabled the creation and sustainment of a larger number of families who could seek a continually rising standard of housing.

Yet such improvements in the quality of accommodation only modified the manner in which new housing was provided.[12] All nineteenth-century provincial cities were typified by a process of development which though capable of prodigious feats of urban expansion in the aggregate, nevertheless was composed of great numbers of small units of production. There was never 'any shortage of speculative builders. Entry into the business required no specialised skills and little or no capital: the smallest of firms was able to co-exist perfectly satisfactorily with the largest.' (Olsen, 1973, p. 334.) In 1870, of the 105 applications for permission to build in the city, 79 (75.2 per cent) were for four houses or fewer and only one was for more than twenty. Potts has calculated that 70 per cent of all building schemes in the city between 1850 and 1900 involved less than five houses (Potts, 1969), and the situation in Leicester was certainly not unusual. The 1851 Census found that two-thirds of master-builders making returns of their numbers of employees had less than ten.[13]

The prosperity of such small builders was precarious. They had to borrow to finance their operations and possessed little capital that was not directly tied up in their products. Such were the vicissitudes of the Victorian economy that they might often find themselves in a situation where their products became unsaleable at a price with which they were able to pay their debts. Building was an important source for investment. Between 1870 and 1914, residential building was responsible for 20 per cent of domestic capital formation (Aldcroft & Richardson, 1969, p. 45) and the proportion was far higher in a very rapidly growing place like Leicester. Investment in property was often an attractive proposition given the growth of urban areas: 'Lending on mortgage was the passive 5 per cent way of taking part' (Dyos & Reeder, 1973, p. 376). In the 1850s and 1860s, the rising standards of housing made the need to mobilise capital more urgent, and it was in this period that one sees the rise of some of the financial organisations that are so familiar a part of today's housing market. There developed in Leicester two major local building societies, the Permanent and the Temperance,[14] a city-wide Freehold Land Society, and a number of special institutions whose aims were to raise capital for specific building projects.

Building societies were founded to lend money on the security of property

(Seymour Price, 1958; Cleary, 1965). Today, that primarily means lending funds to people who wish to become owner-occupiers. In the nineteenth century, this was a minor purpose. Then, building societies generally maintained a much lower percentage on the value of the secured property that they were prepared to advance. One of the most significant roles which the building societies actually performed was to act as one of the means of lubricating the housebuilding market by allowing small builders to borrow fresh funds to continue their operations on the security of those properties which they had just completed but which they might not yet have sold to landlords. Certainly, the records of the Leicester building societies do not indicate any great activity in extending home-ownership. Very few individual householders used them, and of those that did a large proportion seem to have been shopkeepers and publicans.[15]

The character of building societies in this period is reflected in the parallel operations of Freehold Land Societies.[16] Like many of its contemporaries, the Leicester Freehold Land Society was founded with as much of a political as an economic purpose. Its intention was to act as an investment institution which would enable its members to purchase plots of land such that they could qualify as Parliamentary Electors under the franchise of the 1832 Reform Act.[17] Between 1850 and 1879, the Society purchased and developed thirteen separate estates containing a total of nearly 2000 plots. Indeed, so successful as a property development enterprise was the Society that its original intentions were soon completely subordinated to this role. The Society actually reached its peak of operations after the 1867 Reform Act and the extension of the borough franchise to adult male householders had removed its original rationale.

The Freehold Land Society acted as a form of collective speculation in which members took shares, were allotted plots in a ballot and then in practice sold off those plots at a profit to developers or builders. As early as 1855, the Society Annual Report referred (not entirely approvingly one feels) to plots 'changing hands at a premium', and by 1857, the Society was openly advertising its virtues as an investment. The last four estates that it developed in the 1870s and which contained some 900 building plots were certainly constructed in this manner.[18]

Both building and freehold land societies' operations illustrate the predominant characteristics of the Victorian housing market (at least when compared to today). In the nineteenth century, most of the legal, administrative and financial complexities of the market were not so much concerned with the occupation of property as with its ownership, and occupation and ownership were for the most part quite separate factors. The City Rating Valuations for 1855 which include details as to occupiers and owners indicate that only about 4 per cent of all the properties in Leicester were owner-occupied.[19] That the vast majority of properties were let by private landlords gave the housing market at least a superficial unity that today's circumstances lack.

One must condition this statement with the adjective 'superficial' because the processes involved in the operation of the market were extraordinarily complex. Landlords were of much the same character as builders and though there were large landlords (and a man with fifty houses would have been a large landlord in the Leicester of this time), many had only two, three or four houses. It is in this area that least is known: 'Between tenant and landlord often stretched a whole chain of shadowy intermediaries, held in their contracted order by a series of sub-leases which divided responsibilities for the upkeep of the property and inflated the rents paid for it' (Dyos & Reeder, 1973, p. 380). Dyos and Reeder are describing London. Leicester was not quite so involved a situation as this, but did have the same basic characteristics.

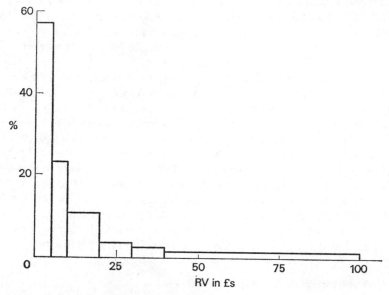

Fig. 3.1. Structure of rateable values in Leicester, 1855

In 1870, Leicester had about 20,000 properties, and annual rentals in the city ranged from two pounds upwards.[20] Fig. 3.1 shows the structure of rateable values in the city in 1855, the last year in the nineteenth century, for which a complete valuation list has survived. In that year, 58 per cent of all properties were rated at less than £5 per annum and only 2.2 per cent at over £40.[21] The developments of the period 1855 to 1870 almost certainly effected some alteration in the value structure at the bottom of the market, such that there was a proportional shift from the category of properties under £5 to those between £5 and £10 as more higher quality properties were constructed in the 1860s. Even so, it is extremely unlikely that by 1870 the proportion of dwellings in the bottom value category had fallen below 50 per cent.

It is very difficult to relate the costs of housing in this period to the incomes of Leicester residents. Contemporary statisticians have suggested some figures for the earnings of adult male workers in the period which may indicate a broad relationship between rents and income. For example, Best quotes hosiery workers as earning between 23 and 25 shillings a week (Best, 1971, pp. 78–99), and from other sources it is known that boot and shoe operatives in Leicester itself were earning between 22 and 30 shillings a week at the same time (Department of Employment, 1971, p. 37). On this basis, one might say that as an approximation housing costs accounted for an average of one-twelfth of the head of household's gross income.[22] This may well have held true over a wide range of working- and middle-class incomes; from general labourers earning about 15 shillings a week and paying perhaps one shilling and sixpence a week in rent, up to the most highly skilled artisan and clerical occupations, earning between 30 and 35 shillings a week and paying rents of between three shillings and sixpence and five shillings a week.

But any such relationship between rent and the adult male worker's wage would be confused by much greater variations in household income. Leicester was fortunate in having a higher than average proportion of women workers. For England and Wales, in 1871, the average proportion of women aged over 15 in work was 26.8 per cent; in Leicester the figure was 40 per cent. Furthermore, whereas nationally nearly half such women workers were in domestic service, only 20 per cent of Leicester's women workers were in this occupation. Both hosiery and footwear employed a high proportion of women. 38 per cent of adult workers in hosiery and 27 per cent of those in footwear were women. Of course, women's wages were much lower than men's (perhaps half was a reasonable average), but even so it was probably important for the improvement of housing conditions in the city of this period, that whilst individual incomes in Leicester may not have been greatly above the national average, household incomes had increased quite markedly in the 1860s.

The spatial structure of Leicester in 1870

Fig. 3.2 illustrates the city of 1870. As a preliminary point it is worth emphasising the small size and consequent high density of the city. There were some 100,000 people living within a total area of less than two and a half square miles. The average radius of the city about its centre[23] was approximately three-quarters of a mile: the average gross residential density was about 100 persons per acre, but in poorer areas where the housing was more tightly packed, this might rise considerably. For example, in the area immediately to the west of Belgrave Gate (area 4 in fig. 3.3, was the poorest in the city at this time), average densities were of the order of 300 to 500 to the acre.[24]

Fig. 3.2 also shows the housing structure of the city in this period. It is based mainly on Rating Valuations, which are the best source available in this period (e.g. Robson, 1966; Gordon, 1971). It has already been pointed out that the Valuations for 1855 are the last to survive intact for Leicester, but

42

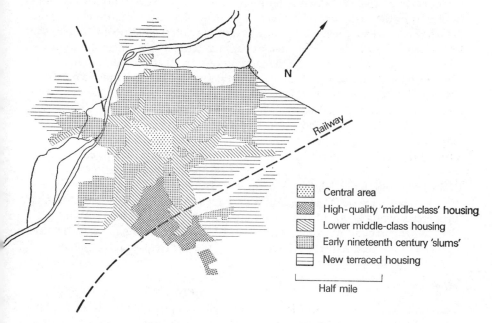

N

Railway

Central area
High-quality 'middle-class' housing
Lower middle-class housing
Early nineteenth century 'slums'
New terraced housing

Half mile

Fig. 3.2. The housing structure of Leicester, 1870

the map does make some attempt to take post-1855 developments into account. Three sources were used for this: (a) those fragments of later Rating Valuations that survive,[25] (b) the 1870 Leicester Trade Protection Society Directory which enabled the location of the higher status areas to be picked out, and (c) morphological information.[26] Fig. 3.2 uses five value-categories, of which a short description follows.

The central area

By 1870, Leicester was beginning to see the emergence of a distinctive, central area characterised by a switch from mixed commercial and residential use of property to a pattern of land-uses dominated by shops and offices. The commercial heart of the eighteenth-century city was in the south-east, especially concentrated about the Market Square, and it was in this district that were situated the homes of the leading businessmen, shopkeepers, merchants and professional men who also used their houses as their places of work. Like many pre-industrial cities, the centre of Leicester saw some functional segregation (e.g. Sjoberg, 1965; Vance, 1967), not only in the case of such trades as the butchers in the Shambles and the dyers on Frog Island, but also such as the solicitors, whose homes were concentrated in New Street, where many of their offices are located today.

43

Housing and the spatial structure of the city

As early as the 1780s, with the laying-out of the New Walk, south-east-wards from the then edge of the city in Welford Place, there had been a trend for the homes and workplaces of business and professional people to become separated. This process accelerated as the city expanded in the first half of the nineteenth century. Patterson has remarked that, 'from about 1840, the shopkeepers, manufacturers and commercial people began to move out from the very heart of the town' (Patterson, 1954, p. 246).

By 1870, this was well advanced in the immediate vicinity of the Market area. St Martins Parish, covering the south-eastern quarter of the historic city, was the only area of Leicester which suffered from a loss of population and falling numbers of dwellings in the middle of the nineteenth century.[27] It is a sure indication of these changes that on the local electoral roll, between 1835 and 1868, whilst the number of residential voters fell, those registered under a business qualification tripled, a measure of the speed with which properties were being abandoned to purely commercial use.[28] What residential property did remain was increasingly that poorer quality housing that had especially resulted from the infilling of the earlier part of the century.

High quality, 'middle-class' housing

The enclosure of the South Field in 1804 opened up the south-eastern perimeter of the city for expansion. The Corporation kept a good deal of this land in its own hands,[29] but the higher ground that extended along the axis of the New Walk was a prime site for good-quality housing. Given the city's problems of drainage and water supply, throughout the period from 1800 to 1870, altitude was a key factor controlling the evolution of the city. The New Walk was the natural corollary to the process by which the city centre had been deserted by the commercial and professional classes who moved into houses built about its axis running from Welford Place to the Old Racecourse. The best property was actually built along the New Walk or in its immediate environs, such as De Montfort Square, whilst at right angles to the main line of development, slightly cheaper, though still very good quality properties were built on the slopes of the ridge. It is sufficient comment on the attractions of this area that even the construction of the Midland Railway in 1840, whose line ran right through the district (albeit for part of the way in a tunnel), appears to have had little influence on the city's social character or pattern of growth (Potts, 1968–9). By 1870, the development of the New Walk was virtually complete. The Corporation's use of a great deal of the land to the south-east blocked off future growth.

As a result, there developed a new style of high-quality housing growth associated with truly suburban locations, physically distinct from the main mass of the city. The classic example, in the case of Leicester, was the Stoney-gate area, beyond Victoria Park, and astride London Road. This district was to emerge as the main concentration of the most expensive housing in the city

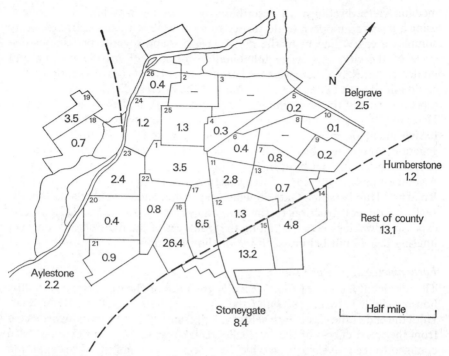

Fig. 3.3. Distribution of persons in the 'Court' directory. Percentage of sample by area: total 979

in the last quarter of the century, but in 1870, Stoneygate was still only the most important of a number of peripheral clusters of expensive properties built in the previous thirty years.

Fig. 3.3 gives a very good impression of the manner in which the residences of these highest income families living in the most expensive properties was changing in this period. It plots the distribution of persons recorded in the 'court' section of the 1870 Leicester directory and as such gives a very good approximation of the distribution of the residences of the best families in the city, the owners of the major businesses, members of the council, leading professionals and so forth. Very few such people were still living in the old central area; only about 10 per cent of those included in this sample were living in parts of the city built before 1800, and a third of the total were resident in the south-eastern sector of the city, astride the New Walk. Over a quarter lived outside the immediate built up area, half of these beyond the inner ring of suburban settlements in the truly rural areas of the county.

'Lower-middle-class' housing

Fig. 3.2 indicates that on either side of the sector especially associated with the most expensive property in the city in 1870, there were flanking blocks of

45

medium value dwellings. In describing these as 'lower-middle class', one is using a term too modern to do justice to what were in reality extraordinarily complex areas. Adjacent to the most expensive and socially prestigious districts of the city, they were inhabited by persons of middle income and moderate status, but there were variations. Such districts did attract those who could afford a bit better than the slums of the poorest districts, and this was especially true of the areas immediately abutting the higher status properties. Here, dwellings were of good quality and occupied by such persons as small businessmen, lower professional occupations and so forth. Further away, and generally lower in altitude, these houses gave way to a cheaper class of property associated with an intermixture of commercial and industrial uses. A good example of this sort of area was that sector of the city between London Road and Humberstone Gate, in which expansion had been blocked off in the 1840s by the construction of the Midland Railway. In such districts, more expensive dwellings occurred along the frontages of the main streets, cheaper housing being built behind such major thoroughfares.

Early nineteenth-century slums

The dominant feature of Fig. 3.2 is the great arc of cheap, very poor quality housing, built between 1800 and the 1850s, in the bend of the River Soar. Until the end of the eighteenth century, Leicester had been generally protected from the worst effects of the river's floods because its built-up area had been confined to the gravel shelf on which had been its original site. The expansion after 1800 soon filled up that shelf, and cheaper housing spread out on to the river's floodplain. Here, the most densely-packed and unhealthiest dwellings were constructed.

That this was the poorest area of the city both in terms of its housing and its inhabitants was universally recognised in this period. The appalling conditions of health and hygiene were acknowledged: 'The district most prejudicial to the health of the inhabitants has been the lower part of the parish of St. Margarets, and particularly the streets and courts to the westward of Wharf Street' (Annual Report of the Leicester Domestic Mission, 1851, p. 12). In this area, for instance were clustered the dwellings of most of Leicester's small Irish colony. Only 1 per cent of the city's population had been born in Ireland according to the 1871 Census, but they and their descendents were predominantly to be found in this district, which had one of the worst reputations in the city.[30] Here, too, at the southern end of Wharf Street were those Common Lodging Houses which provided accommodation for the itinerant population of tramps, peddlers, travelling journeymen and migratory workers who formed so significant a feature of the population of mid-Victorian cities (Hobsbawm, 1964*b*, pp. 34–63; Samuel, 1973, pp. 123–60). In the mid-1850s, it was estimated that as many as 40,000 persons a year passed through the city's thirty-eight Common Lodging Houses (Annual Report of the Leicester Domestic Mission, 1858, p. 15).

In all, about 35 per cent of the city's housing was situated on the low-lying floodplain of the Soar, and this represented the poorest section of the housing market, where properties could be rented for as little as one shilling and six-pence a week by those families with the lowest incomes, most subject to fluctuation. The area was not totally residential, for within it there were a great number of commercial and industrial premises, especially the small workshops associated with both the hosiery and footwear trades. These workshops meant that there were still many small manufacturers living in the area, in or near their places of business, but by 1870, such people were beginn-ing, in growing numbers, to follow the example of the larger manufacturers in deserting the inner city for new areas of housing on the periphery.

New terraced housing

The processes which were encouraging the improving standards of new housing in the 1860s have already been described. Such improvements did not occur overnight, but rather there was a gradual shift towards newer forms of development. The closely packed morphology of the earlier peiod gave way to long terraces, later to be especially associated with developments after the

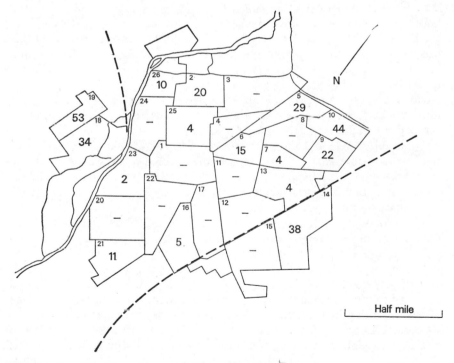

Fig. 3.4. Location of new houses, 1871: number by area

1875 Public Health Act. In Leicester, the first sizeable area to follow the new pattern came with the building over of the Old Cricket Ground on the north side of Humberstone Road, where this small enclave in the midst of the older, poorer-quality houses put up in the 1840s and 1850s was constructed in the new style in the early 1860s.

Fig. 3.4 shows the location of all new houses actually constructed during the single year of 1871. It will be seen that some infilling in older areas is still going on, but by far the greater proportion of new dwellings are being constructed on the periphery of the city. Three main areas stand out.

The first of these was the one associated with the cheapest housing. By 1870, building had all but filled up the quadrilateral of land between the existing north-east edge of the city and the two morphological 'barriers' formed by the Midland Railway and the Willow Brook/Great Northern Railway, and in areas 3, 5 and 9 in 1871 the first houses were beginning to appear across these 'barriers' both on Belgrave and Humberstone Roads.

The second area of such growth was on the west bank of the river. The immediate river bottom was an area of extremely marshy ground subject to almost annual inundations where there was a long-established community of poor boatmen, carters and casual labourers. In the 1860s, terraced housing spread up from these slums to cover the slopes leading to the line of the Fosse Way (now Fosse Road) along which there had grown up one of those little knots of middle-class suburban housing described earlier.

The third of the areas of terraced development which were growing at this time, and almost certainly the one associated with the highest quality housing and persons of greatest social status, was that district to the east of the Midland Railway and north of the existing middle-class properties along London Road. The southern part of this area was known as Highfields, and the more northern parts would evolve later into the great mass of late-nineteenth-century terraces known as Spinney Hills.

These three areas did not represent a single, undifferentiated, quality of new construction. All of them were composed of dwellings better than the slum housing of the previous decades, but there were variations which reflect the much more complex social structure that was being encouraged by economic expansion in the 1860s.

The process by which commercial and professional people had left the central area of the city for new peripheral and then truly suburban locations has already been described as having got under way as early as the beginning of the century. By 1871, other sorts of people were being affected by the same sort of processes. Thus, Ashworth quotes people in Manchester who 'found that the exodus gradually involved people lower in the social scale. One of them, writing in 1871, stated that, besides other members of the middle classes, clerks and warehousemen were moving outwards.' (Ashworth, 1954, p. 19.) In Leicester, a number of groups were following this trend. Primarily, the small manufacturers, especially common in hosiery and boots and shoes, began to move outwards in the 1860s and the medium quality housing in the Spinney Hills area was especially attractive to such persons. Hobsbawn has discussed the concept of an 'aristocracy of labour' in the nineteenth century (Hobsbawm, 1964*b*, pp. 272–315), and some of the occupations that he and others have associated with this definition were distinguished by this trend. Engine drivers, for example, are a classic case of 'aristocrats of labour' and by 1870, they were becoming especially common in the Spinney Hills area to the east of the Midland Railway.

The significance of this fifth category in the city's housing structure was the manner in which it represented a departure in the pattern of residential locations and a big step towards much greater sophistication in social segregation. It is now the task of this study to examine how such changes were being reflected in the pattern of mobility and migration existing in the city in 1870.

Residential mobility in the Leicester of 1870

In seeking a source of information to identify the major characteristics of movement in Leicester in 1870, one must recognise that there are great difficulties. Probably the most useful source for such a study would be the Borough Rating Register, but the last full coverage during the nineteenth century for this source was in 1855; therefore, an alternative was sought in the Borough Electoral Rolls.

In 1870, the status, composition and drawing up of the Borough Electoral Rolls was governed by the provisions laid down by the Reform Act of 1867.[31] In the most simple terms, this had enfranchised all adult male householders who had been living in the area for at least a year, plus any lodgers paying more than £10 per annum in rent.[32] Unfortunately, the actual way in which the Electoral Rolls were constructed was considerably more complicated and is examined in detail in appendix E. For the purposes of this study, the main disadvantage of the Borough Electoral Rolls is that they are not even a comprehensive source for all householders in the city. In 1871, in Leicester, for instance, only about 70 per cent of all householders were on the Roll. There were three main reasons for exclusion: (1) sex – about 15 per cent of all householders were missing because they were women;[33] (2) recent immigrants to the city – because of the residence qualification people who had just arrived in Leicester would have been excluded; and (3) omissions through error, objections to registration or just plain disinterest.[34] Because of these problems, it is very important to stress the need for considerable care in the use of this source, and to emphasise that although broad outlines of the residential mobility process can be obtained from this data, such conclusions need to be interpreted in conjunction with as many other sources of information as possible.

The basic sample for the study was constructed through a comparison of the Electoral Rolls for 1871 and 1872. What was done was that a complete list of all the changes between the two rolls was drawn up by a simple comparison of whether people of the same names were recorded at the same addresses in the two successive years. The total electoral population of Leicester in 1871 was 12,802; that for 1872, 13,337. There was therefore a net gain in the city's electoral population over the year of 3.2 per cent, good evidence of the rapid growth of Leicester in this period.

Given the initial definition of residential mobility made in the introduction, the study is concerned with three types of behaviour – the creation and dissolution of households, movements by established households within the city and movement into and out of the city.

The creation and dissolution of households

Since it is only possible to consider adult male householders in terms of the material available here, the definition of the creation and dissolution of households indicated in the introduction has to be narrowed slightly. In particular, it is simplest to begin by concentrating upon the role of deaths and marriages in initiating changes in the housing market. To do this, permission was obtained from the Registrar-General to examine the death and marriage registers for the Leicester Registration District during this period.[35]

Taking deaths first, the registers were examined to cross-reference with all those persons who were not living at the same address in the 1872 Roll as they had in the 1871.[36] In all, 2725 persons were 'lost' from the 1871 Roll in this fashion, which was 21.3 per cent of that year's total residential electoral population in Leicester. It was possible to trace 457 of these householders through the registration of deaths, i.e. 3.6 per cent of the 1871 electoral population or 16.8 per cent of the 'losses' from that year's Roll.

There are obvious difficulties in trying to check this figure against alternative sources. In 1870, there were 1120 adult deaths in Leicester and 1137 in 1871 (MoH Annual Reports, 1870 & 1871). On this basis, there were about 600 adult male deaths in the city annually in this period, so the deaths of householders are about 75 per cent of the total. It can be estimated that there were between 16,000 and 17,000 male householders in Leicester in 1871 (for other areas see Anderson, 1971; Armstrong, 1974), and they accounted for about three-quarters of the total males in the city aged over twenty (i.e. 23,000). Because the electoral population is smaller than the expected total of adult male householders (only 80 per cent at the most optimistic estimate), the recorded number of 457 deaths is about 100 more than might have been expected, but this is almost certainly explained by the simple fact that the age-distribution of householders (and thereby their death rate) was much older than that for adult males as a whole.[37]

If we turn to marriages, the situation is even more complex.[38] In compiling the list of changes, it was found that there were 3260 'gains' to the 1872 Roll (i.e. persons who were found at addresses at which they had not been recorded in the previous Roll). These 'gains' were 24.4 per cent of the total recorded householders in the 1872 Roll, and 856 of these householders were found to have married in the period before the compilation of the 1872 Roll. These formed 26.3 per cent of all the 'gains'.

Marriage was an important element in the city's migration and mobility system in this period. It is known that on average in the early 1870s, there were about 1000 marriages a year in Leicester, so the total of 'gains' recorded as marriages is a high one considering the character of the electoral population. But this situation is really not so surprising.

Evidence in Leicester suggests most strongly that the sharing of dwellings by two or more families was uncommon. Multi-occupation was very rare in Leicester until after the First World War. Houses in the city were primarily small cottages and although overcrowding did occur, it was mainly through large families rather than the sharing of families. The average household size in 1871 was 4.64, which was almost identical to that in 1901, when there were only 500 families (under 1 per cent) in shared accommodation. Though much of the available housing left a great deal to be desired in terms of quality, there is little evidence of any overall shortage of dwellings.[39] The major reason for this was almost certainly less the virtues of the Victorian housing market than the realistic attitude of those who wished to marry. Since children had to be assumed to be an inevitable consequence of marriage, the prospect of substantially increasing the size of a household without expanding its available accommodation was not one to be faced lightly. This, and the general relationship of cycles of prosperity to marriage, is reflected in the age of marriage. In 1871, only 69 married men in the city were under 20 and the average age of marriage for men was between 27 and 28 years old.[40] As such, it seems reasonable to suppose that the great majority of newly married couples did find a home of their own immediately or very soon after. The bulk of the missing 15 per cent of marriages were primarily made up of re-marriages (almost entirely of widowers since divorce was virtually unknown in this period) which on national evidence should have accounted for well under 10 per cent of all marriages,[41] and people who had got left off the Rolls for the reasons of error, objection and so forth.

In summary, there is no doubt that the natural events of death and marriage were an important factor in the fluidity of the housing market. Given the nature of electoral rolls in this period, one cannot say with certainty that every death represented a vacancy in the housing market. It did not. In many cases, a widow and/or children would have survived, but the scale of the changes indicated is probably a reasonable reflection of the proportion of houses becoming available through the deaths of all householders in this period, i.e. about 4 per cent per annum of the city's total housing stock, and a sixth of all the changes in householders in the year. The excess of marriages over deaths is also a reflection of the importance of the rapidly increasing population of the city in this period and the role played by natural increase as well as movement to Leicester.

Movement within Leicester

Having eliminated the 'natural' elements in the pattern of household change, the two lists of losses from the 1871 Roll and gains to the 1872 were cross-referenced to discover the scale of movement within the city. The process of cross-tabulating the lists by name is obviously fraught with dangers. The possibilities of error are clear, but two precautions were taken. The first was to check wherever possible against the street indices of the Leicester Trade Protection Society Directories for 1870 and 1875. Unfortunately, the proportion of householders recorded in this source is of course a minority and therefore a second precaution was taken by not counting any case in which there were two gains in the 1872 Roll which shared the same name as a loss from the 1871 (or vice versa, with two losses from the 1871 Roll and one gain in the 1872). Obviously, in these cases, there was the possibility of two moves and in the forty-six examples where this occurred, it was decided not to count them as moves within the city (though they remain as a minor contributor to the total rate of residential mobility). Altogether, 1424 internal movers were recorded, which was 11.1 per cent of the 1871 Roll, 52.3 per cent of all the losses from that Roll and 43.7 per cent of the gains to the 1872.

Immigration and emigration

There are therefore two elements left to be considered. The total losses neither recorded as the deaths of householders or moves within the city were 844 or 6.6 per cent of the Electoral Roll for 1871 and 31 per cent of the losses from that Roll. By contrast, the gains to the 1872 Roll so far not explained total 980 or 7.4 per cent of the electoral population and 30.1 per cent of that Roll's gains.

So far as these residual gains and losses are concerned, a large part must have been played by movements to and from Leicester. Between 1861 and 1871, the city's population had experienced a net gain through migration of about 16,000. Assuming that immigrants had approximately the same household size as the indigenous population in 1871, this would have meant an addition of about 300 to 400 households a year through this means. Since this is a net figure, it must be reckoned that the actual number of households arriving in the city in any one year would have been substantially greater. Indeed, it is not impossible to explain all the 992 additions with which we are left through this means, but it must be recognised that there are other possibilities.

One, with which the section on marriages has already dealt, is that couples previously living with their parents or lodging in someone else's dwelling could have set up their own separate establishments, but it has been suggested that this must have been a fairly minor element. Much more likely is that a proportion of these additional gains actually resulted from the complexities of the registration process and the fact that perhaps 10 per cent of all adult male householders were disenfranchised for some reason or other, exclusive of their having recently moved to the city.

Parallel to these processes bringing people on to the Rolls, the same sort of effects were taking them off. Whatever the net effects of immigration in this period, the actual number of families coming to Leicester in any one year was always to some extent counter-balanced by the large numbers who left. Contemporary observers were well aware of the significance of such emigration from the city: 'Whatever increase can have taken place must arise from an influx of persons from other places, but with this there must also be taken into account a great number who from various causes are always migrating from large towns' (MoH Annual Report, 1863, p. 6). Indeed, in the 1850s, when economic conditions in the city were very poor, Leicester actually suffered a net loss of migrants. Even in the late 1860s, it would not seem to have been impossible that many of the otherwise unrecorded losses were due to people leaving the city.

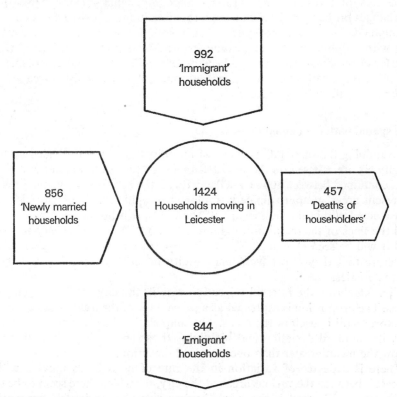

Fig. 3.5. Summary of the system of mobility, 1870–1

It is true that the net gain of 136 households suggested by comparing these unrecorded gains and losses does not correspond to the average annual net increase of households experienced in this period. Of course, the problems of not comparing like with like are very great, and there are again alternative circumstances that could account for such losses of householders.

The overall problems of the Electoral Rolls and their compilation have already been mentioned, and their significance should not be underestimated. But one should also note other reasons which might have caused householders to leave their own accommodation because of economic circumstances. This was especially so for old people, and as well as death, it is possible that some older people were compelled to seek shelter in the workhouse.[42]

Summary

Fig. 3.5 presents a diagrammatic summary of the 'system' of mobility in Leicester in this single year, so far as the Electoral Rolls are able to present evidence for it.[43] In all, about 1 in 5 dwellings probably became vacant in a year. This total was made up of three main components – (1) about a sixth of all changes probably occurred through the dissolution of existing households, through deaths and the movement of survivors out of the housing market to live with children or to enter some institution like the workhouse; (2) a quarter of the vacancies occurred because of families leaving the city; and (3) the most important element, a large number of households moved within Leicester itself.

The spatial pattern of mobility

In examining the spatial dim nsions of mobility, the city was divided up into twenty-six areas based on the sub-divisions used for electoral purposes. Each area contained between 300 and 600 entries on the Electoral Roll, and they do represent areas of approximately similar geographic and demographic scale constructed within the framework of the old parish boundaries in Leicester and the lines of the main thoroughfares. Fig. 3.6 shows the twenty-six areas and it will be seen by comparison with the earlier map of the city's housing structure that they also have some validity in terms of housing and social characteristics.

Fig. 3.6 shows the internal migration rate in the city by these twenty-six areas. The proportion is expressed as a percentage of the total numbers on the Electoral Roll in each of the areas. The map shows the areal average for the city, its standard deviation and highlights those areas with values above or below the mean greater than one standard deviation.

There is a degree of variation in the rate of households moving within Leicester between the various areas of the city, for which there seem to be two main causes. The first is that mobility has a moderate, inverse correlation

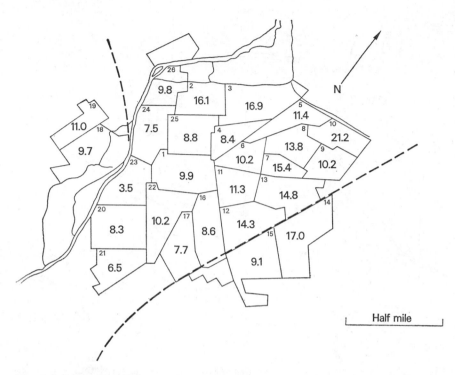

Fig. 3.6. Internal mobility rate by area, 1870–1 (percentage movement by area). $\bar{x} =$ 11.4; SD = 3.9

with social class and the second that newer areas of building have slightly higher rates of mobility on average than the older.

Fig. 3.7 presents the other side of the coin, the proportion of persons gained to the 1872 Roll who had neither married nor moved from somewhere else in Leicester. If one assumes that this is at least a surrogate for immigration to the city, then it suggests that more immigrants (both numerically and proportionally) were coming to the newer areas, especially the cheaper, newer areas, and less to the older and middle-class districts.

Combining these two elements, a simple, general pattern of mobility can be recognised that might be distinguished in terms of five mobility 'regions':

(1) At the outer edge of the city, in the working-class districts of the eastern, northern and western periphery, there were areas of new housing typified by high rates both of movement within and into the city. In these very mobile areas of Leicester, the annual rate of residential change might reach a third of all properties.

(2) Closer in to the city centre from these newer areas, there was that ring of older, early nineteenth century slum districts inhabited by the poorest mem-

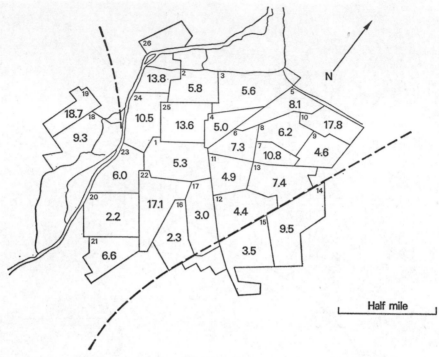

Fig. 3.7. 'Movement to Leicester' by area, 1870–1 (percentage immigration by area). $\bar{x} = 8.0$; SD $= 4.6$

bers of the community living in the cheapest accommodation. These areas also had very high rates of mobility so far as movement within Leicester was concerned but they were far less important than the newer areas in terms of 'immigrants' coming to the city.[44]

(3) The third 'region' of mobility was the historic city centre (excluding perhaps the commercial heart in the south-eastern parish of St Martins (Area 1)). This area had a much lower overall rate of mobility than the first two groups. The rate of movement within Leicester was especially low, but there were more immigrants into this district.

(4) The fourth 'region' is perhaps the most unusual and individual in character. It consists of an area of mixed housing, both cheap and medium-value, immediately to the south of the historic city centre which had grown up very slowly over the period from 1820 to 1870. This was an area of low mobility both within the city and in terms of people coming into Leicester.

(5) The last of the five 'regions' was that formed by the area of middle-class properties strongly segregated in the south-eastern sector of the city. This was that part of the city which had the lowest rates of mobility, both internally and in terms of immigration.

These five regions are general impressions of the pattern of mobility which can only be made more meaningful in terms of the pattern of movement generated within and between them.

The distribution of intra-urban moves

Because the twenty-six areas do vary slightly in their populations, it is more efficient to consider their relationships in terms of movement linkages through percentage rather than numerical measures, especially because of the intention to emphasise the connectivity of the migration system. For each of the twenty-six areas in the city, there will be twenty-seven destinations – itself, twenty-five other areas in the city, and a generalised zone outside the boundary to which some migrants were traced.[45] If every area had an equal amount of movement to all other areas, including those persons who moved within the area, each of the twenty-seven destinations would receive 3.7 per cent of that origin's migrants in the Leicester area. Fig. 3.8 illustrates the pattern of linkages in terms of multiples of that percentage. Thus, the map shows the lowest recorded degree of significant linkage at between 7.4 and 11.1 per cent, i.e. between two and three times the 'expected' proportion: but for moves within each of the origin areas, a much higher threshold has to be used at 22.2 per cent or six times the 'expected'.

This higher threshold for such origin areas is necessitated by the very great proportion of migrants within the city who moved very short distances. In many of the twenty-six areas, over 40 per cent of all such internal moves were actually made within the boundaries of their individual origin area. When one remembers the small size of these areas and the very high rates of mobility indicated, these figures suggest an extraordinarily close mesh of movement, a fact which is reinforced by the great majority of moves which do cross areal boundaries only going into neighbouring districts. However, it is possible on closer examination to recognise elements of a superstructure of intra-urban migration which can be tied in with both the mobility data and the pattern of housing.

A most distinctive feature of this superstructure is in the very poor district of cheap housing, immediately to the north of the historic centre of the city. In this area, it has already been shown that there was a very high rate of mobility within the city, and it can be seen from fig. 3.8 that this mobility was especially associated with very close links within this part of Leicester. The three areas (nos. 2, 3 & 4) which make up this district predominantly are associated with very, very short-distance moves, whose length appears to be constricted by the obvious morphological and housing 'barriers' that surround the district. To the west, there was the river and to the north, the canal. The southern edge was marked by the transition to the rather different character of the historic city. Most surprising perhaps was the little contact that took place across the eastern edge of the district which was formed by Belgrave

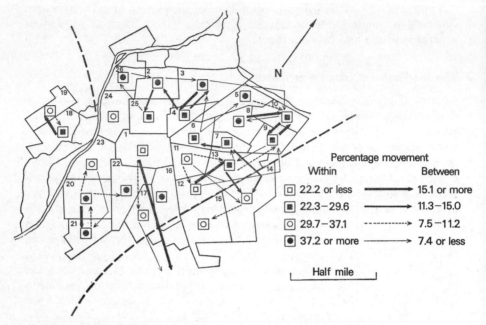

Fig. 3.8. The pattern of linkages by area, 1870–1

Gate. This was surprising both because this was the weakest of the barriers around the district and because the areas to the east were most like this closely interconnected district.

In fact, the large area of poorer housing to the east of Belgrave Road formed the largest and most complex section of the city in which it is possible to recognise close, movement connections. This was not a single simple system of intra-urban migration, but rather a series of interlocked zones covering almost all the parts of Leicester on the northern and eastern sides of the city built between 1830 and 1870. Individual zones can be highlighted in terms of the particular housing and social characteristics that determine their special roles and their directional biases.

Humberstone Road cuts across the whole area and as a preliminary it is interesting that this thoroughfare does not have the same strength as a barrier to movement as Belgrave Gate. To the north of Humberstone Road, the area of poorer quality housing completed in the 1860s does not show as marked a degree of directional bias and has a pattern of linkage rather more like that of the very poor district to the west of Belgrave Gate. But there is some tendency for an outward bias to moves, i.e. to the newer areas (nos. 9 & 10).

South of Humberstone Road, the directional bias is much better developed. One can see from fig. 3.8 how the older areas (nos. 11 & 12) are very strong contributors to the newer districts (nos. 13 & 14) but there is very little back-flow between these districts. The most obvious feature of this area is the morphological barrier provided by the Midland Railway. The railway was probably the strongest and most obvious physical barrier to movement in the city in this period, after the River Soar itself. Yet it seems to have only marginally discouraged movement when compared to the effects of much less definite features like Belgrave Gate. This seems to point to the manner in which such morphological factors can be important in reinforcing existing housing differences but are less able to intervene where there are pressures encouraging families to cross them. The same point is emphasised by the way in which Area 7, the terraced housing of the 1860s on the site of the Old Cricket Ground, is tied more closely to the better quality property to the south of Humberstone Road rather than the poorer standard of dwellings immediately adjacent on its northern edge.

By contrast to the complex movement zone of the working-class and lower-middle-class districts of north and east Leicester, the clearly defined district of middle-class properties in the south-eastern sector of the city forms a very distinct and independent zone of migration displaying a definite pattern of flow. The sector from the old commercial centre in St Martins (Area 1) through the New Walk area of early and mid-nineteenth-century town houses (Area 17) to the new suburb proper of Stoneygate shows a strong outward flow, which even on the basis of one year's information indicates the significance of the pattern of suburbanisation for these middle-class families.

Indeed, it is tempting to see a continuum of zones of migration as one proceeds about the city in a clockwise direction from the older, poorest districts in the north along the river valley, round to the more expensive, middle-class sector of the city in the south-east. Generally, the poorer the area, the more closely intermeshed are its migration flows. High mobility in poor areas is associated with very short-distance moves which have the minimum of directional bias. As one moves up the social scale so this directional emphasis becomes greater because directional bias appears to be closely connected with suburban development. It has been suggested that the process whereby families left the older city centre and moved to newer, better-quality housing on the periphery began in the early part of the century with the removal of the business and professional classes from the city's commercial heart. The pattern of flows in 1871 shows how this was still going on but had now been joined by the further movement of such families away from the main body of the city to truly suburban locations like Stoneygate.

Equally, by the early 1870s, better standards of cheaper housing brought about by a higher quality of demand and the more rigorous enforcement of public health standards were encouraging those with the better paid and more secure jobs also to leave the inner parts of the city to seek newer accommodation on the edges of the built up area. The previous sections of this chapter have demonstrated that the quality of these new, terraced developments of the

1860s was still very varied and it is significant that the directional bias in movement is best developed in that area, Spinney Hills to the east of the Midland Railway, where the standards were highest. The relative weakness of the railway as a barrier to movement, which has been contrasted to the effects of some of the main roads, seems to indicate that many households in this period were eager to take advantage of any new found prosperity to establish themselves in properties quite distinctive from the older, unhealthy districts nearer the city centre.

Over the great arc of housing in north and east Leicester, containing up to two-thirds of all the dwellings in the city, one has a microcosm of the major processes of development that were to occur over the next forty years, as occupational groups, progressively lower down the social order, set out on a centrifugal pattern of movement with the city's growth and the expansion of its housing stock. Because it does reflect the very fundamental processes which were to become of such importance in the city, the migration pattern in this main section of Leicester has been emphasised, but one must also recognise the existence of smaller, subsidiary zones, much more weakly linked by movement to the rest of the city. Three such zones might be recognised.

It is no surprise that one of these should be centred in the area immediately to the south-west of the city centre where the mobility characteristics have been seen to have been different from other poorer areas of Leicester. It is very likely that this was a district 'operating' in a fashion independent of the main bulk of the city's housing.[46] A similar isolated district existed west of the River Soar, though here there was a reproduction of the patterns found in the east of the city. The older area (Area 18) was transferring people to the newer district (Area 19) on the slopes of Fosse Road. Socially and in terms of housing quality, this district west of the river reflects more closely the intermediate areas to the south of Humberstone Road rather than the poor, slum districts. The final zone in the migration system was made up of a number of residual areas grouped around the city centre. These three areas (nos. 23, 24 & 25) do not belong to a mutually interlinked zone, themselves, nor are they greatly connected with any other of the zones that have been suggested in the city. Rather these are heterogeneous districts with no clearly defined character either in terms of housing or population. These inner areas contribute migrants to all parts of the city, though they receive back proportionately fewer people.

It is difficult to determine the degree to which there was any integrated system of linked movements in Leicester in this period. Before any attempt is made to consider the causes of the observed patterns of mobility and movement, it is wise to consider more directly their links with the city's housing market.

Mobility and housing in 1870

In transforming the movement information into the distribution and operation of the housing market, the problems of data considered at the beginning of the chapter loom large. In the market in this period, tenure is not a significant factor; what is important is rental value, since letting totally predominates. It was therefore decided to examine the pattern of movement within and between the broad value-categories derived from the Rating Registers and already used in fig. 3.2. The difficulties of interpolating the assumed values for properties built after 1855 have been referred to, but even so the five categories used in fig. 3.2 do form an approximate structure as good as the sources will allow one to construct.

Table 3.1 shows the matrix of movement flows between the value categories and does reflect the spatial patterns already described. In the lowest value category, i.e. the £0–5 group, there is the greatest proportion of within-group movement. Only just over 10 per cent of the movers originating in this group relocate in a dwelling of a higher value category, and an insignificant number jump up two value categories. It seems certain that the closely knit migration pattern of the very poor districts is reflected in few moves over any great value range and little determinable bias in the net flow of movement.

TABLE 3.1 *Migration flows within Leicester's housing market, 1870–71*

	Rateable value	From					
		A	B	C	D	E	Total
	A £0–5	463	32				495
	B £5–10	76	409	26	3		514
To	C £10–20	4	143	132	18	7	304
	D £20–40		8	25	39	6	78
	E £40 and over				20	16	36
	Total	543	592	183	80	29	1407

By contrast, at the top end of value scale, there is a higher degree of movement between value categories. Of course, it should be pointed out that in the range of values above £20 rateable value, the sort of shifts in costs (and probably in quality) involved in movement are much less significant to the households concerned, since one must assume that at this sort of value, the price of housing has a much looser relationship to income than at the bottom of the market. Nevertheless, the pattern of upward changes in value which the matrix indicates would appear to be consistent with the spatial pattern of relocations from the old city centre to the newer suburban areas both within the city in its south-eastern sector and in those recently developing districts like Stoneygate outside its immediate area.

The most interesting element in the pattern of flows is the movement in and between Categories B and C, i.e. between £5 and £20. The greatest, proportional spillover from one category to another comes from B to C. A quarter of the moves initiated in B end up in C. One can see here the effects of the pattern of suburbanisation spreading down into the lower-middle-class and artisan occupations. Most of the new housing built on the city's periphery in the late 1860s fell into either of these two categories. As a generalisation, the areas to the north of Humberstone Road contained the housing worth less than £10 per annum, that across the west side of the River Soar, a mixture of both categories, and the newly developing Spinney Hills district, the highest proportion of the properties rated above £10 per annum.

The directional bias already observed in all three districts is accurately reflected in the movement within the value categories. One must emphasise that in terms of the total household structure of Leicester in the early 1870s, the strength of this trend towards suburbanisation was not numerically very important. In practice, perhaps two or three hundred households a year were taking part in this movement, at best 2 per cent of all the households in the city. Their importance lay in their obvious relationship to the structural evolution of the city and the clear evidence that they present of the link between the intra-urban migration system and the operations of the urban housing market.

Having said that, it is equally important to emphasise that the predominant characteristic of the total migration picture in the city was its highly mobile population. This mobility was associated with movement both into and within the city. That urban growth throughout the nineteenth century was related to very high levels of rural–urban migration is a very well-known fact, but there was also an extraordinary amount of fluidity within the city, itself.

It was typical of the great bulk of intra-urban moves that they were both short-distance and effected relatively little alteration in a household's position in the housing market. Such a characteristic of the urban movement pattern is not amenable to an explanation in terms of the life-cycle/career-mobility model described in the introduction.

There is no evidence of higher mobility in the better housing areas where the middle-class lived; quite the opposite. The south-eastern sector of the city where the commercial and professional classes had congregated in the mid-nineteenth century was actually the part of Leicester with the lowest rate of mobility, and the districts in the north and east where the slums housed the very poorest were the most mobile.

Of course, career-mobility is not the same thing as saying that people higher up the social scale moved more often. What it says is that physical mobility is closely associated with social mobility and that if a family moves up the social ladder it can be expected to change its address. That is a basic corollary of what has been said about the relationship of housing and social status. The element represented by suburbanisation is typical of what the career-mobility factor should produce, but that was only a small part of the city's migration pattern. In the poorer districts, such a model does not appear very appropriate: here, very high rates of mobility are associated with a spatial and housing flow in which the great majority of moves appear to 'cancel' one another out. In districts like that immediately to the north of the city centre, for every move that is outward or upward in terms of house quality, there is an equivalent that is inward and downward, such that both types of movement are of short distance spatially and socially.

The same limited role in the total movement pattern can be seen for the life-cycle element. Hole & Pountney (1971, p. 15) have examined changes in the life-cycle as part of their general investigation of housing trends over the past century. They demonstrated that the contracting stages of the life-cycle were considerably shorter in the mid-nineteenth century than they are today. Over above and this, of course, was the simple fact that the proportion of householders who survived into what today is termed 'old-age', was so much smaller. In 1871, less than 4 per cent of the total population of Leicester and about 10 per cent of the householders were of retirement age (i.e. over 65 for men and 60 for women). By comparison, in 1971, 20 per cent of the total population of the city and nearly a third of all households was composed of pensioners. It is impossible from this study to state with any conclusive evidence that older people moved less frequently in the mid-nineteenth century city than did younger people. It is possible that they did, but one can say with some confidence that such variation could not have had much influence either upon the pattern of mobility or the structure of the city. There simply were not enough old people.

So what were the factors responsible for the sort of pattern of mobility and migration that has been observed? The most obvious point is that the rate of mobility is highly connected with the frequency of very short-distance moves. The more houses being vacated at any point in time, the greater the opportunities within any given distance for someone seeking accommodation. Therefore, the fundamental question resolves itself into why did people move so often?

Leicester was certainly not unique. Charles Booth could say when speaking of the East End of London in the 1880s, 'The people are always on the move; they shift from part to part like fish in a river.' (Booth, 1889, Pt. I, p. 26.) Twenty years earlier, Dyos noted of Camberwell in the 1860s, 'The impression made on contemporaries, however, was that migration from one street to another, from one district to the next, was on a scale which the tantalising data of the Census never uncovered.' (Dyos, 1961, p. 59.) Even earlier, Razzell quotes very high rates of residential turnover in the 1840s and 1850s in the Aldgate and Whitechapel areas of London (Razzell, 1970), whilst at a smaller scale in the same period, Holmes has suggested that a third of the houses in Ramsgate had a new occupier every two years (Holmes, 1973, p. 249).

Certainly, the mobility of Leicester's population in the middle of the nineteenth century was noticed by Joseph Dare, the Town Missionary. In the late 1850s, he could remark that, 'Many with whom I have formed friendly relations have removed as often as ten, twelve or even fourteen or fifteen times.' (Annual Report of the Leicester Domestic Mission 1858, p. 12.) Dare observed that the migratory habit was, 'sadly destructive of order and progress' (Annual Report of the Leicester Domestic Mission, 1858, p. 12). In this view, Dare was following many social reformers in Victorian cities. Mobility was associated in their minds with poverty, and it is easy to see how frequent movement was forced upon those families whose incomes fluctuated about subsistence level.

For many families, expenditure on housing was closely related to marginal changes in income, in a world, where bereavement, illness or unemployment could reduce even the more prosperous families to abject poverty overnight. In such circumstances, it is hardly surprising that the housing market should have been so volatile. To quote Dare again, 'Some remove immediately on death occurring in the family, some after an attack of sickness, and others borne down by want and poverty vaguely imagine that in some fresh locality, better prospects will open upon them.' (Annual Report of the Leicester Domestic Mission, 1858, p. 14.) The motives suggested by Dare are perhaps a little abstract, and it is simpler to consider that such moves were forced upon householders because of the inevitable need, 'to tap fresh credit, or to find cheaper rooms, or simply to get a "bit decent"' (Dyos, 1961, p. 61).

Most observers were also aware of the predominance of local movement, and generally they stressed as the main reason for this, the importance of finding work. As early as 1842, Chadwick pointed out the difficulties of re-housing people living in slum areas because, 'The workman's "location", as it is termed, is generally governed by his work, near which he must reside. The sort of house, and often the particular house, may be said to be, and usually is, a monopoly.' (Chadwick, 1842, p. 239.) In East London, Charles Booth found that, 'the people usually do not go far, and often cling from generation to generation to one vicinity, almost as if the set of streets which lie there, were an isolated country village' (Booth, 1889, Pt. I, p. 27). Booth could see how specialised occupations in the East End like working in the docks constrained the movement of persons in such poorly paid occupations who had neither the time nor the money to travel far to their workplace. Hobsbawm (1964a) developed these ideas by demonstrating the existence of distinctive regions in the spatial structure of the London labour market at the end of the nineteenth century, and Lees (1970) has demonstrated how a culturally distinctive group, like Irish immigrants, fitted into such a pattern.

The dependence of the poor on a labour market whose radius was described by how far they could walk is easy to imagine as a major control on the pattern of migration within a city the size of London. It is a more tenuous hypothesis in a medium-sized provincial city like Leicester. One should not dismiss it entirely: after all, this was a period in which boot and shoe workers in the city worked a fifty-five hour week and that must have discouraged long journeys to work, and Leicester in 1870 was still a place organised at the scale of the pedestrian. Nineteenth-century cities did perhaps have a layout which made them slightly less convenient to walk across, but on the evidence of the present day, it is very difficult to see how any point in the Leicester of 1870 could have been much further than half an hour's walking distance from any other point. Most of the zones of migration defined in the last section actually had a far smaller radius. The area to the north of the city centre, which had an especially mobile and tightly knit population, for example, comprised a total extent of less than a quarter of a square mile. Within such a 'micro-area', distance alone hardly seems sufficient constraint.

Physical proximity has to be combined with something more specifically related to the association of landlord and tenant in the nineteenth century city. The paucity of information on the direct workings of the Victorian housing market has been stressed, but this lack is in itself a symptom of the conditions facing those seeking accommodation in this period. Most of the means of communication with which we are so familiar today either did not exist or did not exist in the same form in the mid-nineteenth-century city. Sources of information concerning the housing market concentrated on the role of the property owner. Estate agents, rent collectors, solicitors, all the persons working in the housing market, were concerned with ownership rather than occupation. Even local newspapers, which today are perhaps the main source of information for the seeker of accommodation, then concentrated on land and property sales.

Yet this was only one aspect of a more general principle that seems to have been of the greatest importance in urban life, especially the urban life of the poor. Personal knowledge obtained through personal contact was paramount. Wohl (1971, p. 13) has stressed how this applied to the job prospects, social contacts, even creditworthiness of the urban working classes and it is very hard today to visualise just how limited a view of their environment such people possessed or needed. A much greater proportion of one's life could be conducted within a limited area. Indeed, the very high proportion of recent immigrants to the city must have exacerbated this situation. Deprived of the long-term contacts, they must have been especially dependent on those few immediate links that they could build up on arriving in the city.

Such a point also emphasises just how significant in personal terms must have been the movement out to the suburbs being undertaken in this period by an increasing proportion of people in the lower-middle and upper-working classes. Today, a move to the suburbs is often described as an event restricting one's horizons and suburban living is compared unfavourably with the cosmopolitan life of the city centre. In the mid-nineteenth century, the opposite must have been true. A new set of much wider social, home–work and general periphery–centre relationships would have been set up, which at the very least may have encouraged a new attitude towards mobility.

It must be acknowledged that so far only one side of the story has been presented in a demand-oriented framework in which the frequency of mobility has been interpreted in the flexible response of households to changes in their pattern of incomes and the limited radius of movement determined by their need to stay as near to their places of work as possible. This is an attractive hypothesis for explaining the behaviour of the poorer groups in the city, it fits that of the more prosperous less easily. It is true (and certainly not unrelated) that the better-off in the city living in the more expensive houses did move less frequently, but even so, their annual rates of mobility were high by today's standards. Between a fifth and a sixth of all the electoral population changed their address every year. For such movers, the character of the supply side of the housing market was as important as the nature of their demand.

The nineteenth-century housing market was obviously a much better example of 'laissez-faire' in action than that of today. This was a relative rather than an absolute contrast, especially by 1870. Legislative and administrative controls may have been weak and partial but they were extending at this time and having a greater effect on the way the market operated. That the poor should suffer under such conditions, one might almost take for granted, especially given the political impotency of the very poor in this period, but it is also very interesting to consider the manner in which the better-off responded to this situation. In the introduction, the manipulation of externalities has been suggested as one means by which such people could protect their own locational advantages in the city. Yet it is much easier in this period of the nineteenth century up to 1870 to see how the better-off in the city actually took the alternative course of moving their homes away from those areas under the greatest threat from changes in the spatial structure of the city. In this period, one must assume that so rapid was the growth of the city and so ill-equipped the infrastructure of urban life to adapt to that growth that changes in the total equation of locational costs would have been both extraordinarily rapid and peculiarly unsusceptible to any manipulative influence: hence, the way in which the professional and commercial groups left the old city centre for new sites especially suitable for the conditions of the growing city in terms of their competitive advantage for drainage and sanitation.

By 1870, this process was continuing in two ways. Firstly, the centrifugal movement was spreading down the social scale, a process which has been observed and detailed. Secondly, it has also been seen how the richest and most influential families continued their movement away from the city's immediate area to the new suburbs like Stoneygate. This movement might be interpreted in terms of the partial success achieved in manipulating the externalities of the south-eastern sector of the inner city. The crossing of the district by the Midland Railway could not be prevented, only mitigated, and it was not possible to prevent the deterioration of areas bordering that of higher-value housing.

Overall, the lack of tenurial barriers and legal constraints encouraged the market's capacity for adjustment. Changes at one end (e.g. through the construction of, on average, more expensive houses) easily worked themselves through to all parts of the system. Equally, changes at the other end (e.g. the replacement in the central area of residential property by commercial or industrial uses) could soon effect influences up the structure of the market. Rapid adjustments in value and social structure were possible because of frequent mobility. The dominance of short-term renting did not put many obstacles in the way of movement and in many ways encouraged it. Aesthetically, the mid-Victorian city might suggest an atmosphere of monotony, but the very large numbers involved in the supply of housing produced an economic pattern that allowed fine variations in costs and standards. Indeed, the large number of small landlords encouraged flexibility and a subtle response to change if only because they stimulated leads and lags in the market which must have meant continually varying prices, and a width of cost opportunities within a small area, though ultimately, the market must have come close in this period to attaining that sort of long-run equilibrium which it is so difficult to envisage existing today.

Finally, one must mention the most general condition that lay behind the mobility of Leicester's population in this period. This was the natural dynamism of the city. Population growth required a rapid expansion in the housing stock and one must now go on to consider how continuing population growth after 1870 interacted with the changing environment of the housing market to produce the developing system of urban mobility and migration.

CHAPTER 4

HOUSING AND MOBILITY IN LEICESTER, 1870–1940

Leicester's housing market

The system of development that operated between 1870 and the outbreak of the First World War was basically that which had created the city of the first half of the nineteenth century. The city continued to grow as new industries prospered. Fig. 4.1 shows the rate of population growth in Leicester between 1871 and 1971 and it stresses the surge in population between 1871 and 1891. Continued immigration both from the surrounding rural hinterland[1] and from neighbouring industrial towns continued to be of importance.

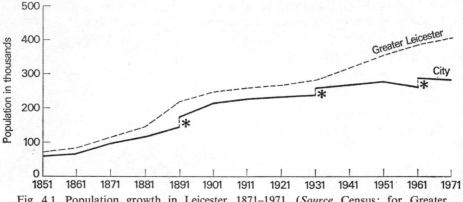

Fig. 4.1. Population growth in Leicester, 1871–1971. (*Source* Census: for Greater Leicester see appendix A.) Boundary changes indicated by *

Jobs multiplied between 1871 and 1911. Hosiery and footwear remained major industries of the city but light engineering increased greatly in these years. The range of industries that Leicester collected has been described as those that no one else wanted, and it might appear curious that the city's prosperity should have been based on trades which are not usually considered as the vanguard of industrial progress. Yet in the late nineteenth century, Leicester was a leader in the shift in the national economy as,

Britain moved towards a new and different economy. It was in large part the economy men wanted, an economy that supplied their needs abundantly and on the whole efficiently. It was able to do this because capital and enterprise themselves responded to the prospects offered by the mass-demand for consumer (and other) goods in a context of rising real incomes. (Wilson, 1965, p. 198.)

The extent to which Leicester shared in this prosperity is amply demonstrated by the famous League of Nations survey in 1920, which claimed that the city was the second richest in Europe.[2]

With a population that rose from 95,000 in 1871 to 227,000 in 1911, an enormous amount of housebuilding took place. In all, about 35,000 new houses were constructed in this period. It has already been seen how there were changes in the standards of new dwellings in the 1860s primarily due to tightening the controls which local authorities had over them. These controls were the result of a rather haphazard course of public health legislation, much of it still specific to particular authorities, and the 1875 Public Health Act is rightly recognised as the major step forward in this field. The Act made mandatory on municipal authorities a range of statutory controls on dwelling densities, ventilation standards, sanitary facilities and the prevention and removal of public health nuisances. Powers were provided for the first time which allowed such authorities to enforce the standards. The effects of the 1875 Act in Leicester have to be interpreted in terms both of the local measures that had taken place in the previous fifteen years and the social changes in the city that the previous chapter suggested provided a much higher level of effective demand for accommodation in the 1860s and was thereby instrumental in encouraging a better quality of house to be built. Throughout the last quarter of the nineteenth century, the driving force which translated the city's growth in population into new housing was the increasing proportion of white-collar and skilled manual workers who were able to afford a rising standard of accommodation.

The apparent uniformity of Victorian cities to present-day observers has already been contrasted with the flexibility with which the housing market operated and the wide range of costs relative to incomes that it provided. Though the basic layout of most new developments up to 1914 remained the terrace, as time went on, there were changes. Johns (1965, pp. 125–6) has said of English cities in this period, 'Street after street displays slight variation in style, and one is compelled to feel that these variations are a direct reflection of the divisions of social class.' Similarly, Forster (1973) has traced how great were the types of building styles that were possible within the basic constraints of the bye-law legislation and has been able to construct a sophisticated classification of such developments in Yorkshire towns that reflects the progress in standards that was made in this period; broadly similar changes may be seen in Leicester.

Such changes were not related to any great alterations in the structure either of the housing market or the process of development but there were necessarily modifications. Primary amongst these was the increase in scale introduced by the needs of a more sophisticated infrastructure. In the early nineteenth century, in a provincial city like Leicester, it had been common for the three functions of 'developer' (the man responsible for the financing of the development, the purchase of the land, the co-ordinating of the various construction activities, etc.), the 'builder' (who put up the houses) and the 'landlord' (who owned the property and leased it to those who actually occupied them) to be combined in the person of one individual.

As the requirements of new housing became more sophisticated, the limited capital resources of the typical small builder could no longer sustain all these roles, and their financing especially became a separate activity. There were a few builders in the city who were capable of undertaking overall, large-scale constructional operations, but they were a minority. Isaac Harrison built a large estate of cottages in the New Found Pool district in the north-west of the city, and Orson Wright carried out a similar scheme in Knighton Fields in the 1880s. Both involved between two and three hundred houses and were especially noteworthy since their builders retained many of the properties so they could act as their own landlords.

What became much more common in the last years of the century were partnerships and companies. Smith, Stone and Donisthorpe, a solicitor and two hosiers, formed the Clarendon Park Land Co., which developed the area to the south of Victoria Park in the 1890s. Other companies included those in the Aylestone Park and Gipsy Lane areas, as well as numerous small partnerships who took part in the development of the Westcotes district. Many of these institutions were formed with a specific purpose and investors were able to take part in a series of schemes as the city expanded. Some individual developers did play a key role. In north-east Leicester, two landowners were important. The Rev. F. A. Burnaby played a considerable part in the development of the Spinney Hills area, whilst the contribution of Arthur Wakerley was unique. Wakerley's North Evington estate was a comprehensive development combining residential and industrial uses built from the mid-1880s onwards (Wakerley, 1913). The formation of such organisations and projects was a reflection of the continuing need to mobilise capital to finance the city's rapid expansion. The institutions that have been described in the previous chapter – building societies, freehold land societies etc. – either changed their character or gave way to a more institutional framework of financing development through joint-stock companies or the normal mechanisms of the banking system.[3]

Housing remained an investment for landlords not for occupiers. The very low proportion of owner-occupied dwellings continued to be a characteristic of the housing market. Fig. 4.2 shows the percentage of dwellings in this tenure in those districts for which the 1895 Rating Valuations survive.[4] Not surprisingly, owner-occupation was concentrated in that south-eastern sector of the city associated with the best quality housing, but even in these areas, owner-occupied properties did not exceed a fifth of the total in any part except at a very small scale. What few Rating Valuation fragments survive for the period between 1895 and 1914 do suggest that the proportion of properties in this tenure may have begun to creep up in the years immediately before the outbreak of the First World War, especially in areas of new housing, such as South Westcotes and North Belgrave. These areas seem to have had more

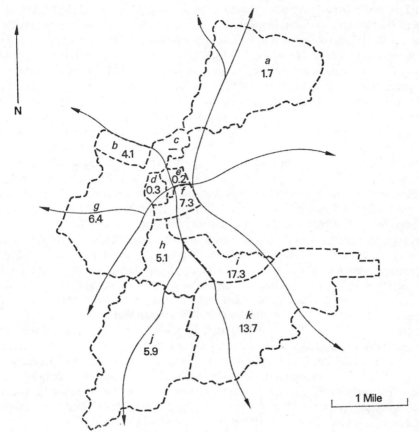

Fig. 4.2. Percentage of owner-occupied houses by Rating Districts. (*Source:* 1895 Revaluation Lists)
Key: *a*. Belgrave. *b*. New Found Pool. *c*. St Leonards. *d*. St Nicholas. *e*. All Saints. *f*. St Martins. *g*. Westcotes. *h*. St Mary's (1): *i*. St Mary's (2). *j*. Aylestone. *k*. Knighton.

owner-occupied properties than would have been true for new areas of a similar social status twenty years before, but even so the proportions involved were still low, and it is extremely unlikely that any area apart from Stoneygate and Knighton had more than 10 per cent of its dwellings held in this manner.

The system of construction and development that operated between 1870 and 1914 had one overriding virtue. It created,

whole suburbs and cities at a rate never before imagined, and it provided houses of all types and sizes at all sorts of prices for a diverse and rapidly growing population. Many of these houses were of poor quality – though they were much better when new than their survivors are today. By the standards of their times, however, they were not too bad; what are now the slums of Manchester and Liverpool were justly described in their day as the finest artisans' dwellings in Europe. (Donnison, 1967, p. 227.)

71

In Leicester, after 1870, such a statement is only partly true. New development though still maintaining a wide range of costs and standards, nevertheless, probably provided for a declining proportion of the city's population, but even so Leicester was one of the system's successes. By 1911, it was one of the least overcrowded of major English cities. The 1911 Census showed an average ratio of persons per room for all County Boroughs in England and Wales of 0.98. In Leicester, at 0.78 that figure was on a par with the least crowded seaside resorts such as Blackpool, Bournemouth or Southport.[5]

Of course, Leicester did benefit from having a larger proportion of its housing stock dating from after the introduction of public health controls, though probably more important was the prosperity to make the controls workable.[6] Yet the city was not characterised by any demographic oddity. There were as many one- or two-person families in Leicester as the average for all CBs in England and Wales, 19.9 per cent. Table 4.1 illustrates the comparison between the average number of persons per family and the average number of persons per building for selected cities in 1911.[7] The table shows just how uncommon was multiple-occupation outside London, Tyneside (which followed the Scottish example) and such ports as Liverpool. Leicester was one of the best examples of a wide class of late-nineteenth-century developed cities dominated by cottages and terraced houses built under the basic provisions of the 1875 Public Health Act and especially found in the Midlands and South of England. The key factor in Leicester was the much greater proportion of reasonably sized dwellings. In 1911, no less than 65 per cent of all families occupied six rooms or more, compared to an average for all CBs of 31 per cent, and only 14 per cent in a really overcrowded city like Newcastle. Only 445 out of the total number of 47,000 occupied residential buildings contained more than one family, and there were only twelve blocks of flats in Leicester with fifty-one families in them.[8]

TABLE 4.1   *Household size and overcrowding, 1911*

City	Av. family size	Av. persons per building
Leicester	4.41	4.46
Nottingham	4.33	4.38
Birmingham	4.70	4.79
Manchester	4.69	4.86
Liverpool	4.91	5.57
London	4.37	7.89
Newcastle	4.80	8.13
All CBs	4.56	4.99
England & Wales	4.51	5.05

It would be totally inaccurate, however, to paint too rosy a picture. There had been enormous progress, but the initial problems posed by industrial and urban development in the first half of the nineteenth century had been as great. The developing public health legislation only applied to new dwellings for the most part, and the powers which local authorities possessed to deal with public health nuisances allowed only the most insanitary houses to be closed. The very legislative actions which were successful in improving the quality of new construction also put such new building outside the reach of the poorer groups in the city. As McKie has pointed out, 'minimum housing standards must be geared to effective demand if there is to be a market in new housing at all levels' (McKie, 1971, p. 18). In the last quarter of the century, such a comprehensive market did not and could not exist. Consequently, all cities were faced with a numerically substantial and spatially segregated stock of old, small, cheap, insanitary and generally sub-standard dwellings, which though declining as a proportion of the total number of houses, still provided accommodation for a great number of their citizens.

Local authorities had obtained powers under Acts in 1868 and 1875 to initiate slum clearance projects (Ensor, 1936, p. 127). These were extended in 1890 but in most cities little was accomplished (Tarn, 1969). Compensation costs put large-scale redevelopment for cheap housing outside the financial capabilities of most local authorities. Leicester Corporation, for example, only built some forty-two municipally owned dwellings under the powers of the 1890 Act. Individual closure orders under the provisions of the public health legislation were more important. Between 1892 and 1914, the Leicester authorities closed some 2000 houses with this power. But usually such demolition was followed by the building of shops and offices, and commercial redevelopment was undoubtedly the main force behind the clearance of older, central-area properties in this period. A special example, much more typical of the early Victorian period in most cities, was the clearance of a large part of Blackfriars Parish to make way for the new station provided for the Great Central Railway's London extension in 1895–7 (Council Minutes, Sept./Oct., 1895; Nock, 1963; Kellett, 1969).

It is probable that there was a considerable shift in public opinion towards some direct intervention in the housing market in the years before the First World War. The 1909 Housing and Town Planning Act had given municipal authorities some greater powers to deal with new schemes of development and there was a parallel move towards dealing with the problem of cheaper housing in central areas. In Leicester itself the Corporation were discussing in 1913–14 the possibility of instituting a much greater direct municipal involvement in the provision of new housing. Of course, any thought of such a scheme disappeared with the outbreak of war in August 1914.

The war was to inaugurate changes in all aspects of British life (Marwick, 1968), but in no sphere was this more true than housing. It brought two immediate difficulties which caused immense strains in the housing market at national and local levels. Firstly, a shortage of labour and materials soon developed as resources were diverted to war production and men went into the Army. Private housebuilding came to almost a complete halt by 1915, and inevitably in cities like Leicester where there was a great growth in war-work, a severe shortage of accommodation developed. By 1916, it could be said that,

At the present time there is a great shortage of houses in Leicester. There are practically no empty houses, and in many houses there are two families. This partly arises from the fact that owing to husbands having been called to the army, wives with or without families have gone to live with parents, sisters or friends. The houses vacated by them have been occupied presumably by newcomers attracted to the town to supply the great demand for labour in the factories. (MoH's Annual Report, 1916, p. 25.)

As well as sharing, of course, overcrowding began to increase. Densities of occupation had been falling steadily in the nineteenth century, from 5.0 in 1871 to 4.5 in 1911 so far as the average number of persons per dwelling was concerned. During the war they rose suddenly and sharply. Whereas only 17.5 per cent of all families in 1911 had occupied four rooms or less, by 1921 this had risen to 27.2 per cent. The number of shared dwellings shot up to 2900 and these were concentrated in the poorest, central area wards, such as Newton and Wyggeston.[9]

The second great problem that the war brought to the housing market was an unprecedented inflation,[10] which combined with the shortage of accommodation to produce a great rise in house rents. As early as the winter of 1914–15, higher rents were causing considerable concern. There were protests on Clydeside and riots in the East End of London. MPs were besieged by the wives of constituents who had volunteered for the Army and now found a soldier's pay insufficient to meet their landlords' demands. The Government decided to take action, and in 1915 passed the Rent and Mortgage Restriction Act which placed a moratorium on rent rises 'for the duration'.

The war completed a transformation in public attitudes towards housing. Lloyd George's slogan of 'Homes fit for heroes' in the immediate post-war 'Coupon' election reflected a concern for the demands in the field of housing which the wartime government had realised would be faced. Wartime committees led to the passing of the 1919 Housing and Town Planning Act (Wilding, 1973). This act provided a generous subsidy to bridge the gap between the costs of building and the rents which poorer families could afford, and thereby encouraged local authorities to embark on much more extensive housing schemes to meet their areas' accommodation problems.

At the end of the war, Leicester Corporation carried out a survey to ascertain the likely demand for new council-provided dwellings. It revealed 1455 households (i.e. 2.6 per cent of the city total) who would like to be rehoused by the local authority. Only 273 of these households did have their own separate accommodation at the time of the survey. The Corporation's housing scheme under the 1919 Act was specifically designed to meet this need, and it was proposed to construct 1500 new dwellings in a programme to run until 1922. The first sites were purchased along Coleman Road, east of the existing Spinney Hills area; in West Humberstone; and in the Knighton Fields area. The scale and purpose of the scheme leaves little doubt that the Corporation did not envisage the 1919 Act as providing for a massive and deliberate intervention in the housing market. Rather,

74

it was a particular solution to the supposedly 'unique' circumstances created by the war.

But it was soon realised that the system of development that has been described as typical of the Victorian period was not being restored. Private building did not recover rapidly after the war and there seemed to have been a loss of confidence on the part of private landlords and private investors. Few envisaged in the early 1920s the fundamental changes that the war had instituted. It was recognised that the continuation of wartime rent restriction policies provided problems but the post-war Coalition Government did not feel able to face the political problems that their complete removal would bring. In 1923, the Onslow Committee certainly thought that rent controls and the fear of such controls were the main reasons why the pre-war development system had failed to re-activate itself (Cullingworth, 1966, p. 18). The Conservative Government of 1923 tried to remedy the situation without too dire political consequences by allowing the decontrol of properties on vacant possession. The 1923 Act also brought to an end the 1919 system of subsidies which had proved far too expensive for central government in the continuing climate of post-war inflation. Only half of the Corporation's 1919 scheme had actually been completed by 1923, but the 1923 Act enabled another 800 houses to be constructed either directly by the council or by private builders with financial assistance from public funds.[11]

The operation of the 1923 Act was shortlived, however, as it was replaced in the next year by the first Labour Government which had come to power. The Act of 1924 (usually known as the Wheatley Act) marks the fundamental break away from the transitional period after the First World War when there was the implicit assumption that arrangements were needed to restore the 'normal' circumstances of the pre-1914 period. The Labour Government had no commitment to resuscitate private landlords and set out to provide a solution to the housing problem through a massive programme of municipal housing.

The 1924 Act certainly brought about a considerable increase in the Corporation's activities. Altogether, some 4700 dwellings were provided under its provisions, being concentrated in three estates: Northfields in West Humberstone; Saffron Lane in Aylestone; and the Braunstone estate on the west side of Narborough Road in the south-west of the city. The Wheatley Act remained in force until 1933, but it was first reinforced and then replaced by the Act of 1930 which was primarily designed to encourage local authorities to take a much greater part in slum clearance. Under this Act, Leicester Corporation cleared much of the oldest (i.e. pre-1845) housing in St Margarets Ward. The displaced population was not rehoused locally in the centre, where the cleared area was devoted to a further expansion of commercial and public activities, but was moved out to the Braunstone estate, where an extension provided another 3000 houses.

75

Between the wars, the total number of publicly assisted dwellings built in Leicester was nearly 10,000, and the Corporation had become an important agent in the housing market. Equally, the private sector of the market underwent great changes. Building to let gave way to building for owner-occupation. The minor trend towards owner-occupation before 1914 has been noted, but the war was again the turning point: 'Higher up the social scale often the only way to obtain a house is to buy one' (MoH's Annual Report, 1918, p. 29). This remark contains a note of shocked surprise that illustrates just how unusual a course of action this was considered. Yet in the inter-war period, far fewer private dwellings were built to let.

The reasons for this were very complicated. Most obviously, there was a much greater availability of finance for home-ownership than there had ever been before. The traditional role of the building societies in supporting the development process through sustaining the activities of entrepreneurs, be they developers, builders or landlords, had to give way to providing money for owner-occupiers.[12] This they did by raising the average proportion of loan against the secured property from 70 to 80 per cent to 90 per cent or more. Furthermore, the whole system received a considerable boost from the fiscal conditions and policies which existed in the so-called 'recovery' period after 1932.

In the 1920s, a low level of private building activity had concentrated on the spread of owner-occupied properties in the upper reaches of the market, where this tenure was already concentrated. After 1932, the situation altered, 'The construction boom of the recovery period was primarily a boom in privately-built unsubsidized residential housing built predominantly for sale rather than to let.' (Richardson, 1965, p. 334.) One of the reasons for the boom was lower mortgage interest rates, but Aldcroft & Richardson (1969, pp. 245–6) suggest that as, if not more, important were falling costs of building (down 10 per cent between 1929 and 1935) and rises in real incomes as the cost of living fell substantially more than wages, which meant that for the employed, a rising standard of living was likely.

In Leicester, the strength and diversity of the city's economy demonstrated itself in the inter-war period by the way in which it stood up to depressed conditions, and many people in the city undoubtedly experienced a very substantial boost to their real incomes. Consequently, owner-occupation was brought within the reach of a much wider range of potential buyers. For those with a steady job (and that meant most in Leicester in the middle 1930s), the possiblity of buying a house on mortgage became a real one. Indeed, in this period, a similar set of circumstances to those described in the 1870s existed with the same sort of status of persons taking the new opportunities of home ownership to move out to a new and higher standard of accommodation.

Fundamentally, the inter-war period saw the collapse of the financial underpinnings of the Victorian housing market and development system.

76

Exclusive of the special problems of rent controls, the economic and fiscal climate provided a far less attractive proposition for investors in housing. Nevitt (1966) has traced how the changing tax position and competitive advantage of rented property as opposed to equities and other forms of investment relatively deteriorated from the end of the First World War. Donnison has summarised the private landlord's changed situation,

> Over the past sixty years, private landlords have been turned into a stagnant and then a dying trade as a result of developments which can be best summarised by saying that the widespread increases in wealth which produced the local capitalism of which they were typical and successful representatives have now gone much further, creating a national and international capitalism that provided formidable competition at every point of the landlords' operations. The same processes have brought about a massive growth in government and correspondingly massive increases in taxes and subsidies which have been organised systematically in a fashion that discriminates against the landlord whilst favouring his competitors. Rent restrictions on which the whole blame for the landlords' difficulties has been cast, furnished additional nails for their coffins – mainly important as a symbol of public discrimination against their trade and as a red herring diverting attention from more fundamental causes of his predicament. (Donnison, 1967, p. 228.)

Consequently, it is not surprising that the 1923 Act had so little success in reversing the decline in building to let. Throughout the inter-war period, the problem of reviving the private-tenanted sector was repeatedly examined. In 1933, as part of an attempt to encourage greater mobility of labour and to reduce the shortage of cheap accommodation, a Rent Act adopted the procedure of decontrolling the most expensive properties, maintaining the middle-range of properties in the condition of the 1923 Act (i.e. decontrol on vacant possession) and stopped decontrol on the cheapest property (which it was thought might soon be cleared in any case). There does seem to be some evidence that the 1933 Act did cause some increase in building to let. Bowley (1944) quotes national statistics that 40 per cent of all housing built by private enterprise in the inter-war period was for letting, but any progress in this direction was halted by the outbreak of war in September 1939 and the reimposition of rent controls on all properties.

Certainly, the housing market in Leicester in 1939 was still dominated by private letting. There are no exact figures for the city, but on the basis of local housebuilding and national evidence, it seems impossible that much less than 70 per cent of all dwellings should have still been in this tenure. But equally, it seems unlikely that there had been any net increase in the number of let properties between the wars. Any new building in the 1930s was off-set by slum clearance and sales of existing housing to sitting tenants. The latter was certainly significant especially in the areas of more expensive properties, where it seems likely that the majority of dwellings were owner-occupied by 1939. In all, owner-occupied housing made up about 20 per cent and local authority property about 10 per cent of all dwellings by this date, and these two tenures had accounted for up to 90 per cent of all the new building that had occurred in the city between the wars.

Both private and public dwellings were built to higher standards than had been generally so before 1914. Terraced forms gave way to much more open layouts centred on semi-detached house forms. Some terraces were constructed in poorer areas, but they were neither as long or as uniform as before the First World War. In the more expensive areas, detached houses continued to be the main form of dwelling. In the public sector, the Tudor Walters Report of 1919 had set new standards of layout and construction to which local authorities should aspire. These were greatly influenced by such pioneer planners as Raymond Unwin, designer of Letchworth and Chief Architect to the Ministry of Health, who had been a wartime advisor on housing to the government (Cherry, 1972; 1974). One only has to compare the open layouts and low densities of the local authority's Braunstone estate of the late 1920s and 1930s with the terraces of South Westcotes, only a quarter of a mile away and an area of very reasonable quality when new, to realise just how great the changes had been.

Because Leicester was a prosperous city in the 1930s, it naturally attracted immigrants from the 'depressed areas' of the North of England. This continued to cause a shortage of accommodation. In the late 1920s, the local authority considered that there were 8000 families in the city requiring a separate dwelling. The 1931 Census indicated that there had been little numerical improvement in the housing situation in the previous decade, despite the local authority building programme, though proportionately the situation had improved marginally. There were still 3000 shared dwellings in the city with 6000 families. Like most local authorities, Leicester gave preference to persons already living in the city, and after the abolition of the Wheatley subsidy in 1933, the main way in which it was possible to obtain a council property was to live in an area liable for slum clearance.

Accommodation problems were lessened by the great expansion of the city's housing stock in the 1930s, but certainly not eliminated. The most obvious way in which excess demand in the city was absorbed was through the continuation of sharing. An alternative was the emergence of a specialised sector of the market associated with the extensive sub-division of properties and a shift towards the furnished letting of accommodation. Lodgings had always existed in the city, either as rooms in private houses or as the special 'common lodging houses' of the nineteenth century. These developments in the 1930s were of rather a different character: multi-occupation was associated with absentee landlords and the splitting up of whole houses into flatlets and bed-sitters. In 1921, there were 112 houses in the city occupied by three families, in 1931 over 200. Most of these were concentrated in the areas of mid-nineteenth-century town houses in the inner south-eastern sector of the city. It was in this district that there seems to have been a significant increase in such multi-occupancy in the 1930s.[13]

The housing market at the outbreak of the Second World War was considerably more complicated than that of 1914 (Bowley, 1944; Cullingworth, 1966). No longer can one recognise any basic unity of operation produced by a single, dominant tenure or a common pattern of response to economic changes. Now, there is not only the very fundamental split between the public and private sectors of development, but the latter is also sub-divided into let and owner-occupied properties, in which there are enormous differences in the conditions of tenure. Overall, the housing market is subject to a far greater degree of administrative control than had been true before the First World War. It is the relationship of these changes in the provision and operation of housing to the spatial organisation and behaviour of the city that will now be examined.

Sources of information

The lack of effective sources of information on mobility and migration is a great problem. Here, the less than satisfactory source of urban directories will be used because of their unique character and suitability for this topic. The limitations on directories are obvious. They are nowhere near a comprehensive source and are biased in the population groups that they cover. But they do have virtues. They are the only source of information about the urban population which contain an indexing system. There are other urban listings going back to the mid-nineteenth century – rate books and Electoral Rolls have been used elsewhere in this study, but in general these sources are much more laborious to use. Warts and all, directories do provide a means of assembling a fair amount of information on the topics with which we are here concerned relatively easily.[14]

The social geography of Leicester

For obvious reasons, the character of the city's directories do not make them very suitable for presenting a comprehensive picture of Leicester's social geography. But there is no other easily accessible source so a number of methods have been utilised here to employ them as fully as possible.

Firstly, it would be useful to consider the character of this source in a little more detail. Fig. 4.3 shows the results of a 5 per cent sample of all the private entries in the set of Leicester directories from 1870 to 1960 at approximately ten-year intervals.[15] The numbers in the directories are compared to the total households in the city as indicated by the relevant Censuses. For most of their history, the directories covered 50 to 60 per cent of the urban householders, only in the 1930s did the proportion rise to over 70 per cent. Directories have always been biased towards the professional and commercial classes, and fig. 4.4 attempts to measure this for the Leicester series by plotting the proportion of persons recorded in them who could be assigned to what one might perhaps term the 'working class'. It is obviously difficult to relate these figures to the overall social structure of the city. Not until the 1931 Census are there any attempts to provide an official Social-Class classification, and it is extremely difficult to transform nineteenth-century classifications of occupations into anything that approximates to the social

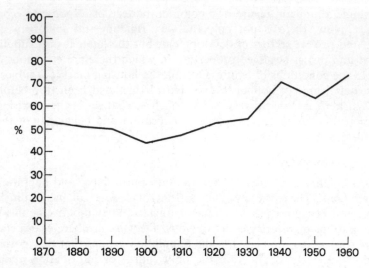

Fig. 4.3. Percentage of Leicester households in directories

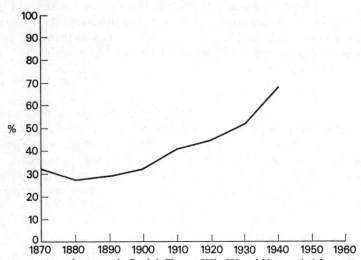

Fig. 4.4. Proportion of persons in Social-Classes IIIb, IV and V, sampled from directories

structure of the city.[16] Present evidence suggests on a somewhat sketchy basis that perhaps 60 to 80 per cent of the city's population (or more accurately its householders) might usually be placed in what today would be described as 'working-class' occupations.

On this rough and ready methodology, one can see that in the nineteenth century the directories can rarely have covered more than a quarter to a third of such persons. Only in the 1930s, when the directories began to include such areas as the local authority housing estates did the proportion of

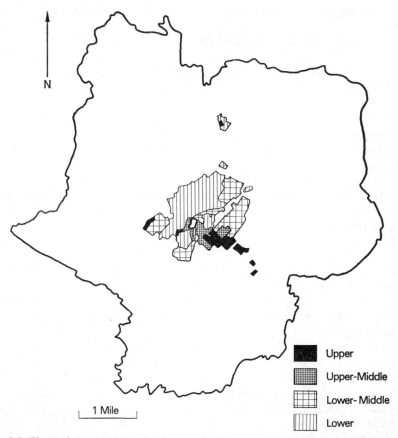

Fig. 4.5. The social geography of Leicester, 1870

working-class households rise to perhaps 70 per cent of the total. Consequently, two general statements about the directories need to be kept in mind.
First, they are biased, but secondly, their coverage of the middle-class groups
is fairly reasonable. The character of the directories automatically determines
the following analysis both in what is possible and in interpreting the results.

As a start, the directories have been used to construct a general overview
of the city's social geography at four points in time – 1870, 1884, 1911 and
1938. For each of the four directories, every street block in the city was
assigned to one of four groupings based on the occupational classification
discussed in appendix C. The four categories are:

(1) Upper – with more than 50 per cent of their recorded householders in
Social-Class I.
(2) Upper-Middle – with more than 50 per cent in Social-Classes I and II.

81

(3) Lower-Middle – with more than 50 per cent in Social-Classes I, II and IIIa.

(4) Lower – all other residential areas.

The potential difficulties are enormous and obvious. There are some redeeming features to the directories, however. Street blocks represented tend either to be virtually completely covered or hardly covered at all. Furthermore, once an area comes into the directory, it is usual for it to remain represented irrespective of the changes that might occur in social status. This is an especially valuable property that will be made use of later. Generally, the placing of all areas not covered in the Lower category is justified by the indcations of fig 4.4, but it must be admitted that this fourfold classification leads to certain oversimplifications so far as 'working-class' areas are concerned.

Fig. 4.5 shows the results of such an analysis from the directory for 1870. The picture painted is one with which the last chapter has made us very familiar. The distinctive sector of middle-class families in the south-eastern sector of the city stands out very clearly and the bulk of the areas in the north and east of the city have been placed in the Lower category. The Upper-Middle group is associated with the fringes of the south-eastern sector and also a number of small pockets in the older parts of the city. The Lower-Middle areas are those recently built in the 1860s, but the distinction between them and the Lower areas in the city is still not spatially well-defined.

Fig. 4.6 shows the city in 1884, and the great expansion of the 1870s is obvious. Virtually all the areas built since 1870 have been placed in the Lower-Middle category and the two features are synonymous with the exception of the better-quality developments to be seen in the south-east and on the western periphery. The 1884 Directory reinforces the suggestion in the last chapter that the cheaper types of new houses built in this period were very much the residences of a mixed group of lower-middle and upper-working-class occupations that were enabled by rising incomes to seek a higher standard of accommodation in this period of the nineteenth century. Typical of the sort of jobs held by the people whom these areas attracted were 'managers', clerks of all types, foremen and numerous skilled manual workers.

Leicester in 1884 was demonstrating two major trends. The first of these was the spatial dispersal associated with growth. Lower densities and the rising costs of new houses encouraged builders and developers to seek sites on cheaper land at the periphery, and the combination of accessibility and the need to connect with such main services as drains and sewers emphasised a radial pattern of development associated wth the main routes out of the city.[17] By 1884, the embryonic fingers of development can be seen clearly stretching out along each of the main roads leading out from the city – Belgrave, Humberstone, Evington, Aylestone and Hinckley Roads.

Complementary to this spatial dispersal was a trend towards spatial segregation produced by the increasing complexity of the housing market. The

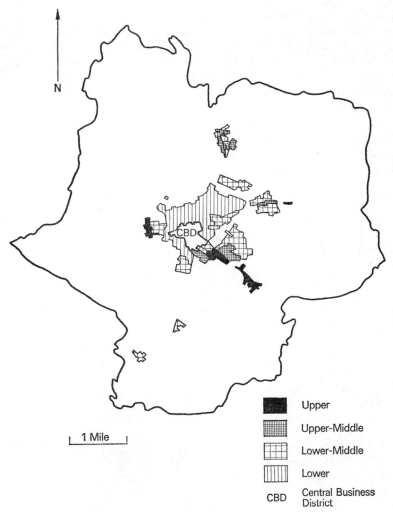

Fig. 4.6. The social geography of Leicester, 1884

most obvious example of this was the growth of the middle-class concentration in the south-eastern sector of the city. The processes observed in 1870 have continued with the city centre being deserted by yet more of these professional and commercial people and the area to the south of Victoria Park growing to make the Stoneygate district a very distinctive feature of the city. But more significant for the greater part of the city had been the increasing disaggregation and dispersal of families whose heads were in those lower-middle and upper-working occupations whom, it has been described, were beginning to move in 1870 in increasing numbers to new properties on the periphery of the city. It is very difficult to present any sort of adequate description of the persons involved in this movement. As a group, they collectively do not easily fit either description of 'middle' or 'working' class. But in 1884, the morphological structure of the city most strongly reinforces the impression from the directories that such a process was going on. In particular, the River Soar and the two railway lines on the east and north of the city now formed an inner triangle which already in the 1880s was becoming associated with the really poor in the city and which divided the sub-standard, cheap housing of the 1840s and 1850s from the better-quality properties of the 1870s and 1880s. Yet it must be emphasised that this should not be seen as a crude dichotomy but rather a simplification

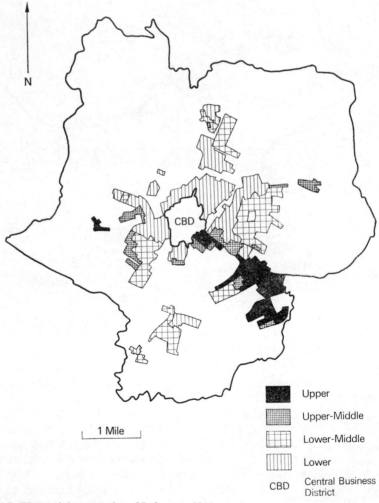

N

CBD

1 Mile

Upper

Upper-Middle

Lower-Middle

Lower

CBD   Central Business District

Fig. 4.7. The social geography of Leicester, 1911

of an emerging and in many ways more subtle segregation of social groups within the city.

Such trends were especially the characteristic product of the late-Victorian city and the climax of the development system that produced this city is seen in fig. 4.7 which illustrates the Leicester of 1911. It is very clear how far the city by the outbreak of the First World War was the product of the same sort of forces and processes that can be traced over the previous forty or fifty years.

The south-eastern sector of the city is now almost completely occupied by areas in the Upper category centred on the fully developed Stoneygate district. Indeed, only small pockets of such people now remain to the north-west of the Midland Railway in the areas which had been the centre of such families' homes in the mid-nineteenth century. Some of the houses in this area have been taken over by the expansion of the central business core, but some have also apparently lost social status and have now been placed in only the Upper-Middle category. Outside the south-eastern sector, only small pockets of such families can be seen in the city, predominantly in the West-cotes and Western Park districts.

Though the central area of the city is now almost entirely dominated by business and commercial land uses, it is closely surrounded and some extent intermingled with 'a very large amount of old and poor class property, consisting of small cottages cheaply built in the first instance many years ago and before the days of modern bye-laws. The rental of these little cottages is generally low and usually from 2/9d to 3/9d a week' (MoH's Annual Report, 1903, p. 61), Within these districts, almost wholly built before 1860, lived the poorest section of the community; 'most of this property is com-paratively near the centre of the town and therefore within walking distance of the place of work – a point of great importance because it is out of the question for persons of this class to afford the penny fares at present charged on the tram-cars' (MoH's Annual Report, 1903, p. 62). Such areas remained a substantial public health hazard. In 1914, infant mortality, for example, in the inner wards of the city was running at three times the rate for the outer, newer, more prosperous ones. In the very worst district of the city (i.e. the poorest and the worst housed), which was generally agreed to be on either side of Belgrave Gate, tuberculosis was endemic and there were still out-breaks of 'fever' nearly every year.[18] In 1911, these very poor districts were almost exactly defined by that inner triangle of land already mentioned in 1884. There were some exceptions like the Upper and Upper-Middle areas to the south-east of the city centre, but the generalisation that this inner triangle held the poorest families and the cheapest housing is a reasonable one. These poor areas contained about 20 per cent of the city's population in 1911.

Out from the central 'slums' (it is the most appropriate term), there had developed and extended that social and rental gradient which was beginning to emerge in 1884. Such a situation was the inevitable outcome of the general improvement in housing standards and rising real incomes that had been characteristic of Leicester in the last thirty years of the nineteenth century, but fig. 4.7 hardly does justice to its complexities. In very general terms, those houses built before 1885 appear to have been occupied by very few persons who can clearly be described as 'middle-class', for example, small businessmen, shopkeepers and so forth, and only a minority of people in non-manual occupations such as clerks and factory foremen. These were the areas which although not considered poverty-stricken by the standards of the time were most clearly and solidly 'working-class'. Typical districts of this kind were South Belgrave, New Found Pool and much of Spinney Hills.

Imperceptibly, such areas shaded out into what contemporary opinion termed 'respectable working-class districts'. These were the suburbs of the 1890s, more accurately described as being occupied by a mixed lower-middle and upper-working-class population. For example, Clarendon Park, to the west of the Stoneygate area, was described as a 'high-class artisan district' (MoH's Annual Report, 1909, p. 24). Similar areas were in North Belgrave and South Westcotes, of which area it could be said in 1902, 'it is noteworthy that it is not necessarily wealth which brings health, but rather the absence of poverty. For Westcotes ward, with a prosperous though hardly wealthy population, but with comparatively little poverty, has the second lowest death rate in the city' (MoH's Annual Report, 1902, p. 20). These outer areas were those which most depended on the growth of accessibility in the city and the development of an extensive public transport system to link them with the main sources of employment which remained concentrated in the central area.[19]

The social geography of the city in 1911 marks the high tide of the Victorian development system. The period of very rapid growth in the population was over, but it had created a city very different from that described in detail in 1870. Between 1870 and 1911, the overall area of the city had expanded well over five times and this expansion had been accompanied by great changes in the distribution of housing types and social groups. Yet the city of 1911 was a relatively simple form which might be interpreted in terms of a small number of easily understood concepts – rises in real incomes, improvements in housing standards and accessibility combined to provide a pattern which bears a strong resemblance to the spatial models suggested by such as the ecologists.

Yet if there was this simplicity in 1911, it should be contrasted with the changes that occurred in the inter-war period. Fig. 4.8 illustrates how the building boom of the 1930s in particular encouraged a further outward spread of the city. Again in the inter-war period, lower densities of layouts ensured that the radial extension of the city would be rapid. The southeastern sector of the city remained the location of the best housing, but the Upper groups within the boundary of Leicester have not spread greatly since 1911. Indeed, the properties to the north-west of Victoria Park have now almost totally moved out of this category, and many of them had in fact

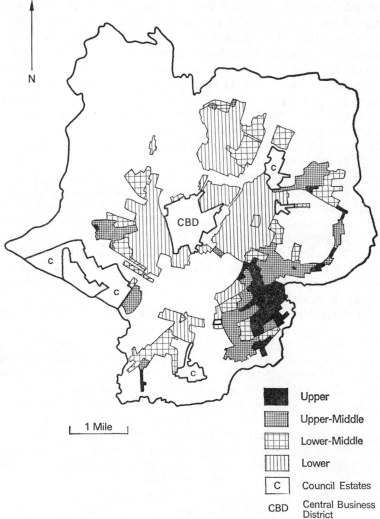

N

1 Mile

■ Upper

▦ Upper-Middle

▦ Lower-Middle

▥ Lower

C Council Estates

CBD Central Business District

CBD

Fig. 4.8. The social geography of Leicester, 1938. 'C' indicates council estates

been converted into multi-occupation following the trend that has already been noted as important in the 1930s. Most of the new dwellings for the Upper category families were not built in the city in this period, but rather outside its limits, either in the natural extension of their traditional location in the south eastern sector, across the city boundary in Oadby or completely away from the city in outlying villages like Desford or Rothley.

The 1930s especially saw the development of great blocks of owner-occupied properties all round the periphery of the city. Some of these areas have been placed in the Upper-Middle category, some in the Lower-Middle. It is not always easy to discriminate as to which of the categories these areas do fit. The former category is typical of areas especially found in Knighton and Evington, on either flank of the south-eastern, Upper category sector, and also on a smaller scale in West Humberstone, in Western Park, in the south-west of the city along Narborough Road, and filling some interstices in the Stoneygate area. These districts were associated with the larger properties that were built in bigger plots. Semi-detached houses were still common, but there were also some detached properties. It is likely that it was in these areas that owner-occupation totally dominated.

But it is also possible to recognise areas of slightly lower social status, which were built in the 1930s and which attracted many of those people who found in that period the opportunity to invest in a home of their own. These areas have a social composition which is very similar to that of the cheaper districts, like South Westcotes and North Belgrave which were built in the early 1900s. They were located in much the same areas as the more expensive developments, with which they are closely intermingled, West Humberstone, Belgrave and Aylestone Park. It is likely that in these cheaper private areas of the inter-war period, some of the houses were not owner-occupied but built for letting though it is difficult now to state exactly what proportion might have taken this tenure.

Undoubtedly, the novel feature of the social geography of 1938 was the presence of the municipal housing estates. Their locations were basically similar to those of the new private estates in that they occupied peripheral sites. This is not surprising since at this time, the local authority was competing as one among many persons and organisations operating in the urban housing market, not as they did after 1945, when the structure of planning legislation gave the local authority rather greater control over the assignment of land. The estates built in the 1920s were on a comparatively modest scale, equivalent to the private schemes of the same period. The Braunstone estate of the 1930s, built in the south-west of the city, is a much larger unit and prefigures the scale of land development that the local authority would be able to undertake after the Second World War.

So far as the inner city is concerned, most of the private housing built before 1914 and outside the south-eastern sector, has now been placed in the Lower category. Evidence for such areas in this period between the wars is somewhat vague, but what is available suggests that those districts built in the last two decades before the First World War lost the few middle-class occupations that had been resident in them before 1914. It would be foolish not to recognise that the Lower category of 1938 must disguise considerable social variation. The social structure of those areas close in to the city centre, which with the demolition of the St Margarets district, had become the oldest, cheapest and poorest parts of the city, was quite distinct from that in such areas as South Westcotes, which were only forty years old in 1938. Yet even so, the limited evidence of the directories does suggest already in that year that a significant difference could be observed between the pre- and post-First World War cities.

This very cursory introduction to the social geography of the city between 1870 and 1938 indicates certain significant points. Most importantly, one must be aware of the speed with which social changes seem to take place in small areas, which is reflected in the way in which they have passed from one category to another. Generally, these changes have been downwards, but even from the limited indications of the directories, it is clear that the speed with which such changes have occurred have varied enormously. To investigate this further, ten sample areas in the city will be examined to see how such local social changes tie in with changes in both mobility and migration and the structure of the housing market.

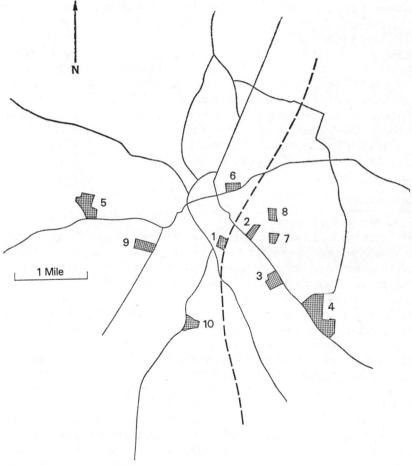

Fig. 4.9. The ten sample areas

Mobility and changes in the social geography at the local level

The ten areas that have been selected for analysis in this section do not represent any sort of statistically valid sample. They are simply ten areas, built between the middle of the nineteenth century and the outbreak of the First World War (with only very limited building in them after 1919), which have been picked out from the directory to represent as wide a range as possible of the areas which are represented in it in detail. All of them were still occupied in the middle 1960s when the cross-sectional analysis of the next chapter was carried out. The areas, though they of course reflect the biases of the directories as a source of information, nevertheless also make use of that property, already mentioned, whereby if an area once comes into reasonably full coverage in the series, there is a high probability that it will remain so covered irrespective of the changes in its social structure that may take place. There is therefore a good chance to examine the validity of such concepts as 'filtering' and 'invasion–succession' over a long period of time. The ten areas chosen are illustrated in fig. 4.9: they all contain less than 200 dwellings and their boundaries have been approximately drawn to coincide with some sort of morphological and housing homogeneity. In the most simple terms, they can be divided into two groups of five, one group that may be termed the 'upper' group since it contains areas originally built for persons in middle-class occupations, and a second, the 'lower' group, built originally for persons lower down the social scale.

The ten areas

1. The oldest of the 'upper' areas is in that part of the city, known as South-fields, that lies to the south-east of the city centre. Today (1974), it is among the oldest inhabited districts in the city. The properties in general consist of mid-Victorian town houses, built between 1844 and 1870. A terraced layout predominates, but when new the houses were sufficiently expensive to retain some individual characteristics. This area lies just to the south of the New Walk, and although it never contained the very best housing in the city even when brand-new, the standard of dwellings was high and the district remained an attractive one with a high proportion of professional and commercial families right into the inter-war period. But in the 1930s, this was one of the districts most affected by the trend towards multi-occupation and the splitting up of larger houses into furnished tenancies. Such a process was well advanced by 1939, and today there are very few, if any, of the dwellings in this area which are still in single family occupancy. Such changes have been accompanied by a deterioration of the local environment; many houses are poorly maintained; the streets are unkempt and the district does not have an attractive local reputation.[20]

2. In many ways, Area 2 is very similar to Area 1. It is slightly younger, but also consists of mid-Victorian town houses. As with Area 1, it retained much of its original character up to the First World War. The main difference was created by London Road, a major traffic and commercial artery of the city, which forms its southern boundary. There has been a constant and continuing pressure down this road for the redevelopment of residential properties to business uses. Most of the houses along the frontage of the main road have been so converted with their upper floors becoming offices or in some cases flats. Again, as with Area 1, multi-occupancy began to spread in this area before the war and is now almost totally dominant in those dwellings which remain in residential use. In the 1950s, this was one of the districts into which many of the city's first wave of Commonwealth immigrants came, and it remains an area of transients. Like Area 1, these social changes have been accompanied by a physical deterioration.

3. The third area represents the next stage in building for persons whom the last section placed in the 'upper' category of areas. This area was built up mainly in the 1880s, with some later infilling, and lies just to the south of Victoria Park, on the west side of London Road. The area marks an important stage in the transition of development to lower densities and more open layouts. Houses were no longer mainly built in terraces, but as individual dwellings, detached in their own grounds.[21] Multi-occupancy did begin to appear in this area in the late 1950s, especially when the expansion of the University brought more students looking for accommodation nearby. But usually conversions have been of a higher quality and the area retains an attractive appearance. Many dwellings remain in single-family use, though in recent years there has been a growing tendency for developers to look to areas such as this as a source of sites which can be redeveloped to provide high-quality, high-density accommodation near the city centre. A number of sites have been cleared of their original houses and new blocks of 'luxury flats' erected.[22]

4. Area 4 marks the last stage in the outward spread of housing built for people in the 'upper' category from the city centre to the Borough boundary. This area lies right in the south-eastern corner of Leicester, in a triangle between London and Stoughton Roads and the city boundary. First building took place here just after the turn of the century, but the process of development was rather more drawn out than in the other three areas. Plots were built up singly but development was interrupted by the First World War and some construction went on into the 1930s. In the years immediately before 1914, this was probably the most expensive area in the city in terms of house values and the most exclusive in terms of social status. It has remained an area characterised by upper-middle-class families and multi-occupancy is still rare. One of the trends of recent years, however, has been the infiltration of many more institutional uses such as hostels, old peoples' homes and so forth which have taken over certain of the large houses.

5. The last of the areas that have been distinguished as 'upper' in category

is slightly different from the other four in that it is on the opposite side of the city from the south-eastern sector in which the others lie. This fifth area is in the west, in the district known as Western Park, between Hinckley and Glenfield Roads. Again here, the first development occurred before 1914, but building was interrupted by the First World War. Prior to the war, development was of a high class of good-quality, detached dwellings, but after 1919, there was a good deal of infilling of cheaper property. Social change was much more rapid in this area in the inter-war period and many of the houses soon passed into multi-occupancy in the years after 1945.

6. The sixth area is the first of the second set of districts which represent that housing built by private enterprise before the First World War for people lower down the social scale than the families in the middle-class who first occupied the houses represented by the first five areas. Like that first set, the second five areas are arranged chronologically to cover the period of development between 1860 and the First World War. The first of them is that district to the north of Humberstone Road built on the site of the Old Cricket Ground, which has already been mentioned in the previous chapter. The houses here were built before the 1875 Public Health Act but were one of the first large-scale terraced developments in the city. The area was more expensive and of a higher social standing than the older districts which surrounded it, and therefore it survived longer, and when it was finally demolished in the early 1970s was one of the poorest and most delapidated areas of Leicester.

7. This is the next stage of the late-nineteenth-century development process and represents one of the earlier schemes in the Spinney Hills district of the city. It is unusual because it was one of the Leicester Freehold Land Society's estates developed in the 1870s, but this does not seem to have had any great influence in making this area atypical of the general pattern development in this period.[23]

8. This area is about half a mile north of Area 7 and very similar in morphology and the type of house it contains. Like Area 7, it is typical of much of the cheaper developments of the late nineteenth century in having generally larger and more expensive houses fronting on to the main roads such as Evington and Melbourne Roads. These larger houses have been especially susceptible to changes since the First World War and many of them have been converted to shops, or such facilities as doctors' or dentists' surgeries, and of course more recently to multi-occupancy. Equally, both Areas 7 and 8 have been centres for the location of Commonwealth immigrants, especially from India, Pakistan and East Africa, in the last fifteen years.

9. This area is on the other side of the city to Areas 7 and 8, and is about twenty years later in its age of development. It is situated in South Westcotes and is representative of that last great burst of building that occurred in the city before the First World War. Most of the houses date from the turn of the century and are morphologically distinctive from those of twenty years before. Plots were larger, house design allowed for greater individuality through

such features as porches and bay windows, and the area has a much more open appearance because of the front gardens that the houses possess.

10. The last area is rather different from the other four in this set but still typical of an important type of development in the second half of the nineteenth century. The first developments in this area came with the spreading out from the city centre of those upper-middle-class commercial and professional occupations in the middle of the nineteenth century. Most of these families moved to houses in the south-eastern sector of the city but some moved elsewhere to isolated groups of high-cost houses constructed in peripheral locations, usually along arterial routes out from the city. One of these groups was constructed along Aylestone Road and behind it in the 1880s there grew up a group of small cottages often containing people serving the local community. Such nuclei were common in many areas of the city in this period – Belgrave, Aylestone, Humberstone. Originally, they were quite separate from the main mass of development, in this case the open land of Freeman's Common intervened, but as more and more housing was built so the areas were swallowed into the bulk of Leicester. Their service centres grew as small nodes of local convenience shops, but large developers found their vicinity inhibiting for their activities and in the inter-war period they often represented loose areas of building with opportunities for infilling. After 1945, when development conditions on the urban periphery have been much tighter owing to planning controls, these opportunities have often been taken up. As a result, this area is unusual in having a far less homogeneous morphology with a wide mix of house types and ages of building in close juxtaposition.

In looking at changes in these ten areas, three simple criteria will be used: social change as measured by alterations in the occupational composition of householders; changes in age-structure so far as they can be traced from the limited information of the directories; and lastly, variations in the rate of turnover of dwellings in the areas.

The evolution of the sample areas: (1) social change

Fig. 4.10 shows changes in the occupational structure of householders in each of the ten areas, using a sixfold classification, together with the proportion of women householders with no occupation and the properties taken over for commercial use.[24]

Fig. 4.11 collapses each of the diagrams into a set of simple indices constructed by allocating six points to every householder in Social-Class I, five to every one in Social-Class II, and so on down to one point for every householder in Social-Class V. At each decade, the total score of the area is summed and divided by the number of occupied householders to provide an average. For convenience, the division of the ten areas into an 'upper' and 'lower' group will be continued.

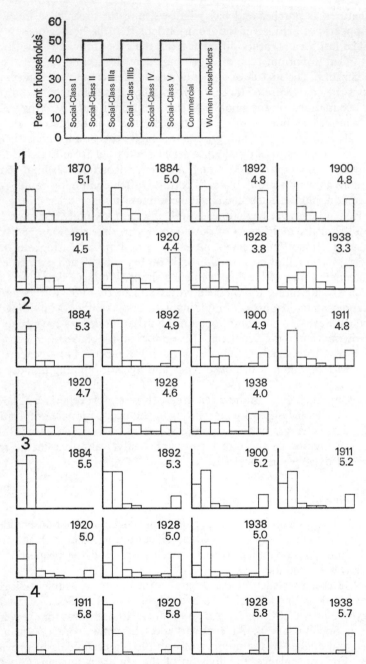

Fig. 4.10. The ten sample areas: social change

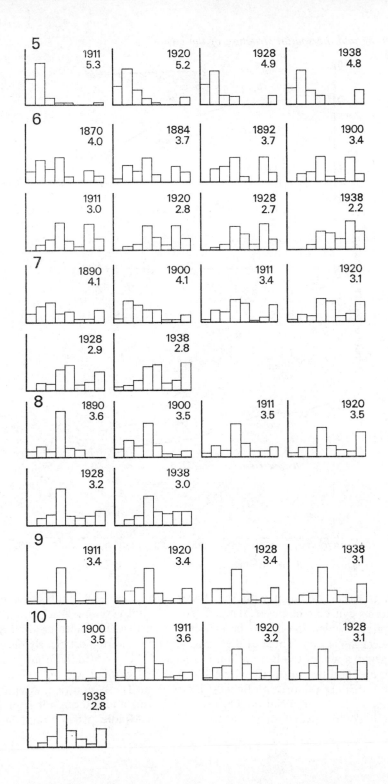

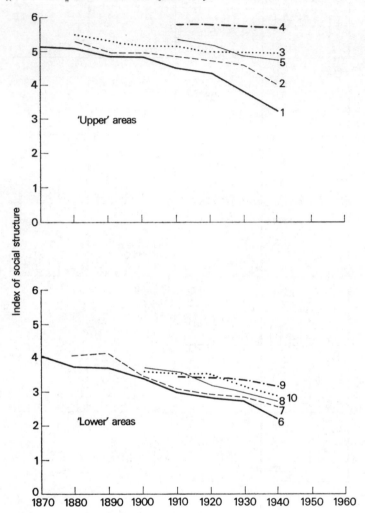

Fig. 4.11. Summary of the social change indices

It is easier to begin with the 'lower' group as very similar residential histories can be recognised. There is some initial variation in the quality of the areas, which is only to be expected even in the conditions of the late nineteenth century. Three of the areas, 6, 7 and 9 were most clearly occupied by persons whom the last section has termed 'Lower-Middle', that is to say with a concentration in the clerical and skilled manual worker groups, but with a distinct minority in Social-Classes I and II. By contrast, Area 8 was of distinctly lower social quality even when new and although it is possible that it could have been classified as 'Lower-Middle' when brand new, it

96

must soon have fallen into the 'Lower' category with very few persons in Social-Classes I, II or IIIa. Area 10 is slightly different, for the reasons of its different history of development already described. Its slow growth meant that when the first decennial count from the directories was taken in 1900, a few of its cheaper properties must have been twenty years old, and it does seem that its mix of housing types is reflected in a wider and less distinctive social structure.

However, the generalisation that when new all these areas were inhabited by a mixed lower-middle and upper-working-class population still holds true. Though the proportions varied slightly, none of these areas could accurately be described as poor. Indeed, it is very characteristic of them that they have a relatively wide range of occupational types when new – from skilled artisans in such trades as building and engineering, through clerks, foremen and managers up to small businessmen and even some professional occupations. It is equally true however that this middle-class component did not stay in the areas very long. The time scale for this loss obviously varies with the size of the original middle-class component in the population. It has already been described how Area 8 soon lost the bulk of these middle-class occupations, except for the small component that continued to live in the larger houses fronting the main thoroughfares, and by 1900, it was a predominantly artisan district with its remaining lower-middle-class inhabitants more than balanced by a growing minority of semi-skilled and unskilled workers.

The same sort of pattern characterises Areas 6 and 7, though they start from a 'higher' initial point and change proceeds slightly more slowly. It took about twenty years for Area 6 to lose its middle-class inhabitants and about twenty-five years for the same to happen in Area 7. It is overwhelmingly clear that in occupational terms, these 'lower' areas have always contained a fair degree of spread about any social mean and that although changes in social structure occur, they do so slowly and without any apparent resemblance to the sort of stepped progression from one social structure to another characteristic of the invasion–succession model.

It also seems clear that it is possible for this sort of progression to go on for a very long time, and for areas to sink gradually down the social scale, and one must also assume down the housing market's value structure. One may take Area 6 as the longest example. Built by 1865, in 1880, it could still marginally be described as a 'Lower-Middle' district thanks to the numbers of small manufacturers living in this area near their places of work. By 1911, these middle-class households had declined to the point where there were very few of them indeed, and the numbers of non-manual workers of all kinds were in a minority. The numbers of semi-skilled workers had grown but even in 1911, the properties were probably too expensive for the poorest families. Artisans were still the majority. It was in the inter-war period, with the demolition of the St Margarets area that this district became one of the cheapest in the city and one sees how semi-skilled and unskilled workers in fact grow towards a majority as the social structure becomes more and more heavily biased towards the poorest groups.

It can easily be seen in terms of the progress which the other areas make along the same path that the process tends to repeat itself at different points in time. Thus, Area 9 seems to follow about thirty years behind Area 6. In 1911, it was still hovering at the margins of the 'Lower-Middle' category, but during the 1920s and 1930s slowly slipped, losing middle-class families and gaining presumably less well-off, working-class ones.

In most ways, the 'upper' areas do not display a different form of change, though they do initially represent a much higher level of housing quality and social status. All five areas when new were inhabited by a firmly middle-class population, though again there was some variation. Area 4 was clearly a very high status district with a very large proportion of its inhabitants in Social-Class I. The other districts are perhaps more accurately described in terms of the categories of the previous section as 'Upper-Middle', with a mixture of families whose occupations were in Social-Class I and II. Perhaps, social change in these better-off areas was somewhat slower than in the first set of five. For example, Area 1 in 1920 when it was nearly sixty years old, was still very obviously middle-class in its population composition with concentrations of managers, professional men and the owners of small businesses: the same was true of Area 2 on the other side of London Road. Not only do changes in occupational structure occur slowly but such changes do not appear to fit any cyclical pattern of residential evolution. All the areas demonstrate that it is very rare for the housing market to juxtapose persons of radically different social standing so that the degree of perception required of residents in recognising the long-term processes at work is greater than most people can manage. But these upper areas also illustrate the sort of circumstances that may arise to trigger such a level of perception. In particular, the inter-war period saw a much more rapid pattern of change produced by the development of multi-occupation. This is most obvious in Areas 1 and 2. Area 1 was still middle-class in 1920 with few working-class householders, but in the 1920s, there was a very rapid rise in the numbers of skilled workmen resident and by 1938 it was no longer a middle-class area by any standards. The greatest fall in the index used to plot changes in social class occurs in this period and the same is true of Area 2 where the situation was complicated (though probably reinforced) by the parallel spread of commercial properties.

That these 'upper' areas do display some different characteristics in their patterns of residential change from the 'lower' areas suggests a number of points related to what has been said in the introduction. The 'lower' areas quite clearly display a 'normal' pattern of evolution, if 'normal' is thought to be similar to the model of 'filtering' changes in the housing market. It was pointed out in the introduction that 'filtering' is more strictly reserved for value changes rather than for changes in the social structure of areas, but that the two processes are (so far as the long term is concerned) very closely connected. It is typical of residential changes in the 'lower' areas that they occur in a way which suggests the gradual and continuous progress of 'filtering' rather than the stop/start, lead/lag changes associated with the invasion–succession model.

By contrast, the 'upper' areas immediately suggest the strength of the forces operating against such changes. It is likely that the relatively slow processes of occupational change in these middle-class areas represents some degree of success by the inhabitants in manipulating the forces of the market to prevent the social (and therefore the value) deterioration of their neighbourhoods. This is very much the argument presented in the discussion of externalities and the need for group action to influence the character of the housing market. It is an attractive argument for such areas since they are inhabited by those persons who might be expected to have the most influence in this direction, especially perhaps in the political and economic circumstances of the late nineteenth century. It has been suggested that there were attempts to influence the pattern of land uses in the south-eastern sector of the city. But equally, one must recognise that the very character of the housing market from the mid-nineteenth century onwards has encouraged slow social changes in such high-status areas. It has seemed likely both in this and the previous section that the dwellings designed for the groups at the top of the occupational scale and the housing market were fundamentally different from the bulk of the new houses, which themselves were originally built for only a minority at the upper end of these distributions. They were different because they were larger and morphologically distinctive, but more particularly, they were 'protected' in housing terms from social change by the buffer provided by the new dwellings built for families in the 'lower-middle class'. These families were the natural inheritors of the better quality dwellings, but in fact they seem to have preferred to improve their residential environments by seeking new dwellings, of better standards, built in the next accretion of growth at the city's periphery, a trend excellently demonstrated by the overall pictures of the city's social geography provided by the directories in 1884, 1911 and 1938. Some of these persons did filter into the better houses but the bulk of this population was diverted by alternative accommodation in terms of the outward spread of the cheaper types of new housing.

In these terms, it is possible to see how the distribution of the best dwellings in Leicester is the result of a combination of processes very similar to that described by Hoyt (1939) and discussed in other contexts both in Australia (Johnston, 1966) and Britain (Richardson, Vipond & Furbey, 1974). Most of the familiar features are here – the sectoral form, the attraction of high ground, the clustering together of community leaders, the slow accretion of growth at the periphery and the gradual decline of the high-rent/high-status area. What further suggests itself is the degree to which the character of the housing market itself reinforces the existence and stability of such a form.

It has been characteristic of Leicester that the highest value housing has been protected because it has formed such a small proportion of the new building in any period of the city's recent history and because the groups next in line to take over such houses in terms of the filtering model have actually improved their housing conditions by seeking more modern dwellings further out on the city's periphery. It seems likely that the more the city grew and the longer the time that passed so the attractiveness of the larger, high value housing declined for the middle-class families who might have been expected to occupy it because such larger properties became increasingly obsolescent and unsuitable for modern living. But if this was so, why did they remain for so long associated with the better-off families? The answer may well lie in a look at the other two factors to be considered in this investigation at the local level.

The evolution of the sample areas: (2) age structure

However scanty and obscure the information on the occupational structure of such small areas, that relevant to their age-structure is almost non-existent. Fig. 4.12 is an attempt to use what suggests itself as a possible measure of ageing in the populations of such areas – the proportion of householders recorded as married women. For much of this period, it is probably safe to assume that the bulk of these women were widows and as a general principle, widows should be at the older end of the demographic spectrum.[25]

Two caveats need to be mentioned. The first is that taking the proportion of widows as a measure of the age-structure is a relative rather than an absolute standard. Given the overall changes in age-structure produced by the greater longevity of the population, the widows of 1870 almost certainly were, on average, considerably younger than those of today. Secondly, one needs to consider whether there has been any general rise in the proportion of householders who are widows. In 1951, the Census suggested that 12.1 per cent of all households were headed by widows and another 4.4 per cent by single women. Hole and Pountney (1971, p. 21) thought that in 1851 about 17 per cent of all households were headed by widowed and divorced persons (though that of course includes widowers and male divorcees). Such evidence would indicate that a major change in the numbers of house-holders who were widows has been unlikely.

Thus, though this is a rough and ready estimate, the profiles for the proportions of widows acting as householders do provide certain general indications. As a start, the 'upper' and 'lower' sets of areas display rather different patterns. The 'upper' set of areas have consistently higher levels of widows as householders than the 'lower'. This variation may well be explained simply by the greater ability in the nineteenth century of widows in the middle classes to sustain their own households.

It is true of both sets of areas that they display a tendency for the proportion

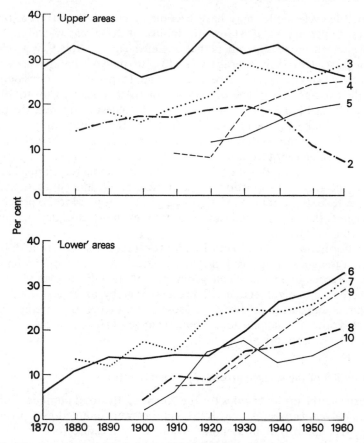

Fig. 4.12. The ten sample areas: widows as householders

of widows to increase as the areas age physically. All have very few widows when new, which suggests a not unexpected correlation between recently built areas and younger age groups. Then as the areas grow older so they seem to age demographically.[26] This is some evidence for a life-cycle control on mobility at least so far as it seems to have affected the evolution of the areas but one cannot make too much of it because the 'young' period for all areas is so short and there seems little evidence prior to 1914 of a general life-cycle control on the social geography.

After 1920, this is less true because the process of ageing in the areas seems to become more severe, and this is so for both 'upper' and 'lower' areas. If there has been no basic change in the proportion of widows in the city, this would indicate that these areas increasingly moved into a position where they were older than the city's average. It also argues that the age-structure of sub-areas within the city became more differentiated and as a corollary

101

that the life-cycle model may have become of greater importance. In 1911, the average percentage of widows in the five 'lower' areas was about 11 per cent, which when one remembers that one of the areas, Area 9, was relatively new, suggests that their average was close to that of the city as a whole, especially when the variation between 8 and 15 per cent is considered. By 1938, the average percentage of widows for these areas was up to 19.6 and variation from 14 to 26 per cent. In Area 6, the oldest of the five, the proportion nearly doubled from 14 to 26 per cent. For the 'upper' sample areas, the graphs point to a more complex picture. There is not so consistent a pattern, and in the nineteenth century there was usually a higher proportion of widows in these areas. But there is a trend towards an increase after 1920, and by 1938, the average proportion of widows as householders in these five areas had risen to 20.9 per cent. Variation was greater, from 14 to 32 per cent and this suggests that 'upper' and 'lower' areas grew more alike demographically in the inter-war period.

The information of the directories on age-structure is very limited, but one does gain the general impression that age-structure has become a feature of greater importance in the social geography of the city. It also seems likely that there is a general and usual process whereby areas are inhabited by older populations as their dwellings grow older. Before any attempt is made to fit this into the conclusions about changes in occupational structure, turnover rates have to be examined.

The evolution of the sample areas: (3) turnover rates

In measuring the turnover rates for the ten areas, the total number of changes in householders between successive directories were counted as a proportion of the total number of households recorded by the directory in that area. Unfortunately, one has to take into consideration here the irregular time interval between the publication of directories. From 1870 to 1941, this interval varies from two to six years. It is obviously necessary to try to reduce the varying proportions to some common base, preferably an estimate of the turnover per annum. The statistical problem involved is a complex one and has received some attention (Cave, 1969). Here the situation is confused by doubts about the exact accuracy of pinpointing directory periods to dates of publication.[27] Consistency is probably as sufficient a virtue as can be hoped for. Therefore, the relatively simple compound interest formula,

$$Pn = T(1 - r^n)$$

was used, where Pn is the proportion of the population still living at the same address after n years, T is the total number of households in the district at the beginning of the period under consideration and r is the annual rate of turnover. Fig. 4.13 shows the turnover rates for the ten areas, again split into 'upper' and 'lower' groups. It is most useful to consider the pattern of turnover rates in terms of two elements – a secular and cyclical trend.

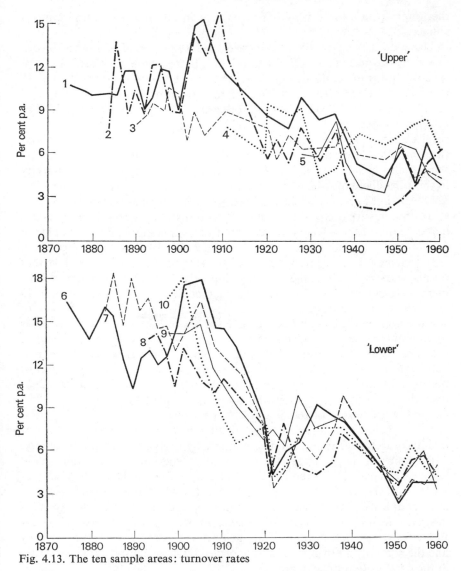

Fig. 4.13. The ten sample areas: turnover rates

For the upper set, the secular trend seems well developed with a fall from an average rate of turnover in the late nineteenth century of between 12 and 15 per cent per annum down to 3 to 6 per cent in the 1930s. The trend is similar for the lower set with a fall from 14 to 18 per cent per annum before the First World War to perhaps 6 to 8 per cent in the 1930s. Equally, both sets of areas display cyclical variations. It must be admitted that to some extent deficiencies in the data can magnify these but the swings, especially in

103

the lower set, are too severe and prominent in all areas to be the result of chance.

There does seem to be a well-developed sequence of ups and downs, especially obvious in the period before 1914. Before the First World War, the turnover rates in the lower set of areas were characteristically higher than those in the upper areas, and the inverse correlation between mobility and social class suggested for 1870 is reproduced. Although the First World War brought changes in the behaviour of both sets of areas, one of its main results was to pull down mobility in the poorer housing areas to levels about the same if not lower than those of the better. Thus, Areas 3 and 4, the highest social status areas of the ten, and amongst the highest status areas in the city in this period, are those least affected by the war. Indeed, it is hardly a noticeable event in these areas' residential histories other than as one step down in a steady progressive decline in mobility. By contrast, the poorest housing areas, 6 and 7, are those on which the affects of the war were quite catastrophic in producing an enormous fall in turnover from which they have never recovered.

In terms of the general evolution of the sample areas, it seems clear that one is witnessing a complex relationship involving the interaction of the overall development of the housing market and the relative changes in property valuations that that produces with the parallel redistribution of social groups in the city and the relative success or failure of individual areas to prevent the effects of such changes. Throughout the late nineteenth century, social change in the lower set of areas was a continuous process associated with high rates of mobility. During this period, there is very little evidence of the residents of areas in the lower set being able to exert much influence over the pattern of changes in the total equation of locational costs. The continuing expansion of the city meant that there was an unremitting pressure from below to which they responded by moving out to newer and better quality property which was almost continually being built on the periphery.

By contrast, in the upper set of areas, the development of the spatial patterns associated with the Hoyt model can be seen as the result of a combination of the structure of the housing market which diverted pressure into such peripheral growth, the technical obsolescence which increasingly characterised the properties inhabited by such groups, the lack of need or inclination to develop alternative uses for such properties such as multi-occupancy and the generally more important role played by such features as age-structure in the composition of such areas.

Changes in age-structure and the life-cycle

Any study of such residential areas as has been attempted in the previous section cannot hope to form a totally comprehensive picture of the development of the city's social geography. The areas themselves are insufficiently reliable a sample, and the data which have been applied to them too uncertain in character. But certain general hypotheses are beginning to emerge which need to be followed up. As with the pattern of turnover rates, these hypotheses suggest a division into secular and cyclical explanations.

The secular explanations centre on the apparent long-term fall in mobility, the evidence of a change in the relationship of mobility to social class, and the general question of the manner in which such changes might relate to the particular development of the life-cycle model and the role of age-structure in the city's social geography. One way of looking at these factors in more detail is to consider changes in the residential mobility behaviour of certain occupational groups. No single occupation can be considered to be sufficiently representative of any particular social grouping to act as a surrogate. Two sample sets of occupations were selected to approximately fit the sort of persons who first occupied the 'upper' and 'lower' sample areas, and they will also be referred to as 'upper' and 'lower' samples.

The 'upper' sample parallels the idea of an urban elite as used by Johnston in his studies in Melbourne and Christchurch (Johnston, 1966; 1969a). The concept of an elite in a metropolitan city like Melbourne is rather different from that which applies in a medium-sized city in the provinces like Leicester. Here, an elite might be defined as those persons who one can expect to be able to exert the greatest influence on the development of the urban area, either because they hold direct roles as landowners, large builders, controllers of sources of local finance, members of the public authorities and so forth, or indirectly as large employers and significant members of the local business and professional community. Five occupations were chosen as a basis for this upper sample: four professions – solicitors, doctors, accountants and architects; and a general commercial group – company directors. The first four are fairly self-evident. So far as company directors are concerned, the spread of the public, limited liability company in the mid-nineteenth century meant that this group plays a more and more important part in the composition of the sample.[28] They are especially useful in distinguishing the more important members of semi-professional groups who have achieved a higher status over the last century and whose incomes may depend on the size of their practices (e.g. estate agents).

The 'lower' group is far more selective. It represents those in the community whom it is supposed have generally been able to maintain a reasonably competitive position in the housing market, whilst perhaps lacking the sorts of controlling influence that may be ascribed to the upper sample. There is only one professional group in the lower sample – teachers, and four commercial – managers (who were not directors), commercial travellers, builders and shopkeepers with a home address separate from that of their businesses (thereby eliminating the smaller shopkeepers).

Both sample groups are most fairly described as middle-class (if that term has any real meaning), if only because they are not among the poorer groups of the city. They do not include such occupations as clerks and factory foremen who might be claimed to represent the next step down on the social ladder.

One cannot underestimate the difficulties involved in interpreting occupational data as a surrogate for social class, but it is the only realistic source available.[29] Over as long a period as seventy years, such problems are magnified. There must be changes in the social ranking of occupations, both within the two sample groups and the social structure as a whole. Even the nominal descriptions of occupations can change. For example, commercial travellers (or persons who might admit to that description) have probably declined in status in relation to managers. Many professions have attained corporate status and have been able to erect institutional barriers to entry to achieve higher salaries and greater social ranking, though for instance teachers have not gained as much success as accountants or architects.

The two sets of sample occupations were drawn from the same directories that have already provided the basis for the general maps of the city's social geography. Three directories were used – 1884, 1911 and 1938. A large sampling fraction of 1 in 4 was taken at these relatively infrequent intervals

TABLE 4.2 *Occupational composition of the samples*

Upper	Lower
1884	*1884*
11 Solicitors	109 Managers
9 Doctors	47 Commercial travellers
6 Architects	54 Shopkeepers
22 Other professional	50 Teachers
32 Hosiery manufacturers	149 Miscellaneous
44 Boot and shoe manufacturers	
9 Elastic web manufacturers	
10 Other manufacturers	
57 Commercial occupations	
55 Miscellaneous	
1911	*1911*
19 Solicitors	160 Managers
28 Doctors	128 Commercial travellers
13 Architects	90 Shopkeepers
37 Other professional	83 Teachers
23 Hosiery manufacturers	205 Miscellaneous
40 Boot and shoe manufacturers	
73 Other manufacturers	
50 Commercial occupations	
52 Miscellaneous	
1938	*1938*
23 Solicitors	257 Managers
40 Doctors	206 Commercial travellers
20 Architects	62 Shopkeepers
60 Other professional	59 Teachers
30 Hosiery manufacturers	181 Miscellaneous
34 Boot and shoe manufacturers	
61 Commercial occupations	
142 Miscellaneous	

because these three directories give exceptionally good occupational coverage, and it was wished to ensure as much consistency in the definitions used as possible. Table 4.2 gives details of the occupational representation of the samples. Both were subsequently reorganised by tracing backwards and forwards in the directories at approximate ten-year intervals (i.e. 1875, 1884, 1892, 1902, 1911, 1920, 1928, 1938 and 1950), and the addresses at which members of the samples were living at these dates recorded. Fig. 4.14 shows

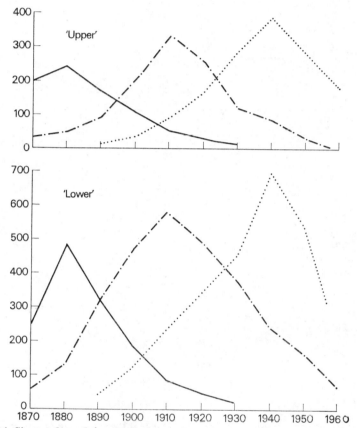

Fig. 4.14. Size profiles of the sample groups

the consequent size profiles of the samples. As one would expect, these are approximate 'normal' distributions, peaked at the year of sampling. In any year, the total proportion of the occupational groups represented in the samples will be somewhat greater than 25 per cent, and though there must be variations from year to year, on average about 40 per cent of the occupations chosen will be included in the samples.

Fig. 4.15 shows the rates of residential mobility for the two sets, using

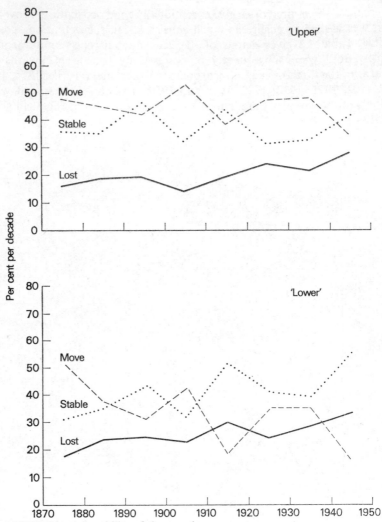

Fig. 4.15. Residential mobility of the sample groups

three categories – stable (i.e. at the same address at subsequent dates), moving within the city, and lost from the directory.[30] Because the same source has been used here as for the investigation of the sample areas, it is not surprising that the same general conclusions may be drawn. More interesting is the difference between the two samples. The fall in mobility is much less marked for the upper sample and indeed these groups have only a marginal decline in rates of movement. Whereas for the lower sample, increasing stability is associated with an equal and continuous process of falling mobility, stability in the upper sample has not increased markedly since the First World

108

War. Rather falling mobility within the city has been associated with a rising proportion of persons lost from the directory.

Fig. 4.16 compares the mobility rates for the two samples with that from the sample drawn from the directories as a whole.[31] Whilst the lower sample displays a pattern very similar to that seen in the total sample, the upper sample is quite distinctive. One of the reasons for this is the occupational relationships implied by the two samples. The upper occupations all would today occur within the Registrar-General's Social-Class I and the lower is equivalent to the sort of people who are placed in Social-Class II. But the

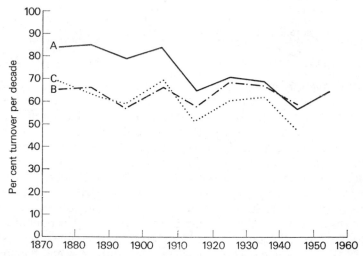

Fig. 4.16. Mobility of the samples compared to total mobility in the city
Key: A. total. B. 'upper'. C. 'lower'.

upper sample is a much higher proportion of the potential 'population' from which it has been selected than is the lower of its 'population'. In a city like Leicester, the 1931 and 1951 Censuses suggest that the relationship between Social-Classes I and II in the total population is of the order 1 to 6, but the equivalent ratio between the two samples is only 1 to 1.6. Given the general importance of the middle class in the directories, this all suggests that there may have been a distinct difference in the migration behaviour of the very best families in the city when compared to the bulk of the middle class. This may be examined in more detail in terms of the two separate elements of the life-cycle and career-mobility model.

Because directories allow one to look only into the behaviour of house-holders, changes in the age-structure of the total population are less important than in those of the former alone. Though both elements must be closely related, the changes in the age-structure of the total population have been more significant. Table 4.3 compares the age-structure of Leicester's total population in 1871 and 1951, and illustrates the marked if not unexpected changes. The ageing of the city's population has to be related to the general fact that between 1901 and 1951, the birth rate fell by 45 per cent, the death rate by 27 per cent and that the level of infant mortality is at present at about 12 per cent of the levels of the beginning of the century.

TABLE 4.3 *Percentage of Leicester's population in major age-groups*

	1871	1951
0–14	35.3	20.0
15–24	20.6	12.5
25–44	26.7	29.6
45–retirement	13.6	23.9
Post-retirement	3.8	14.0

Source: 1871 and 1951 Censuses. Post-retirement is 65 for males and 60 for females.

The age-structure of householders is more difficult to estimate, especially for 1871. Table 4.4 approximates the distribution of their age-structure for the same two years. For 1871, it assumes that all married men and widowers were householders and that the residual distribution of householders approximates to the age-distribution for widows in the city.[32] The 1951 distribution is taken from that year's Census which does allow the construction of such an estimate of the age-structure of householders. This certainly reinforces the impression given from the study of changes at the small area scale that age-structure has become a more important determining factor in the city's geography since the First World War. The doubling in the proportion of older householders is likely both to reflect the growing significance of the life-cycle model in considerations of mobility and the presence of so many older householders ought to have been a factor in encouraging lower levels of mobility in the city.[33]

TABLE 4.4 *Percentage of householders in major age-groups in Leicester*

	1871	1951
Under 40	48·5	26·6
40–59	36·0	43·9
Over 60	15·5	29·5

This can be further stressed through looking at the behaviour of the two sample groups. Each sample set has been reorganised such that it is now divided into 'entry-groups',[34] in which all persons first appearing in the directories in the same decennial period are considered together, irrespective of the directory from which they were originally sampled. The character of the directories and the samples which have been drawn from them means that one cannot say with complete certainty that each 'entry-group' represents a common range of ages. Given the variable ages at which people set up independent households, the importance of immigration to Leicester, and the overall character of the directories themselves, that is impossible, but two properties are useful. The first is that the mean age of each successive entry group at any single point in time is likely to be lower than those preceding it (because it is assumed that they will on the whole be younger), and secondly, one can say with certainty that the mean age of each entry group must rise through time.

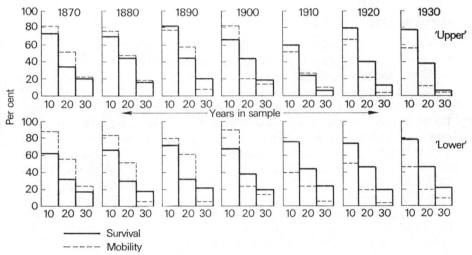

Fig. 4.17. Entry groups, survival and mobility

Fig. 4.17 illustrates two features derived from the entry groups. The first is that it shows the manner in which the average length of time which householders survive in the directories grows longer. As well as householders surviving to a greater age, this may reflect a decline in immigration to the city and a higher proportion of households recruited locally at an earlier age. It is striking that the increase in the period of time which households are present in the directories is much more marked for the lower sample than the upper. Again, as with mobility in general, the behaviour of the upper sample shows very little change over the whole period from the nineteenth century.[35] The second property of the graphs is that they illustrate what proportion of each entry group moved in successive decades. They demonstrate how, irrespective of the changes in the proportion of persons surviving longer, mobility within all age-groups has fallen steadily since the middle of the last century. Similarly, the subsidiary points of much less change for the upper sample and more significant alterations since the First World War emerge.

Given all the limitations of the samples, the main conclusions that may be drawn are twofold. The first is that the most simple assumption of the life-cycle model, that mobility declines with age, has been true over the whole period under consideration, and that therefore changes in the rates of mobility have occurred in 'parallel'. That is, not only have younger people moved less but so have the more elderly. At the same time, general changes in the age-structure of the whole population and of householders in particular have meant that this fall in mobility has been magnified by increasing the proportion of less mobile, older households at the expense of the more mobile, younger ones.

Social mobility

It is extremely difficult to make even the most generalised statements about the career or social-mobility element in the residential history of Leicester with any confidence. Social mobility is an enormous imponderable in this work like in any other long-term study of urban areas. Lampard (1968–9) has made the most significant point that whilst in the United States a central theme of urban studies has consistently been 'the idea of the city as a social escalator', in Britain, the concern has been far more with concepts of class and society.[36] Obviously, in a study concerned with movement and change, a static conception of the city is inappropriate, and the initial premise of studying housing through the process of residential mobility assumes some sort of basic relationship between geographic and social movement.

The methodological problems of examining social mobility are enormous for a sociologist concentrating on this topic; in a study of this kind they are virtually insurmountable. For a start, one is not concerned with the extremes, the rags to riches idea, but with the small movements of income and status that may play so significant a part in the evolution of an urban area. Equally, the information upon which one has to depend, the occupational data of the directories, is fraught with difficulties and dangers.

Fig. 4.18 illustrates an attempt to measure a small part of the social-mobility process as it appears to have influenced the behaviour of the two sample

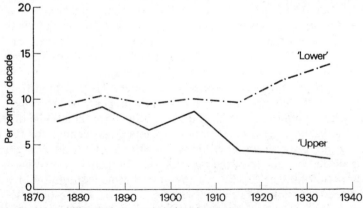

Fig. 4.18. Social mobility and the occupational samples

112

groups. Though these samples were selected and defined on the basis of certain occupations, the process of tracing people backwards and forwards through the directories produced a number of cases where people apparently changed their occupations or sometimes the titles with which their occupations were described.[37] Many of the problems associated with the comparison of names from the electoral rolls are faced here, and given that changes of occupation are often associated with changes of address (which is after all one of the foundations of this study), there are obvious grounds for error.[38]

Fig. 4.18 shows for each occupational set of samples, the proportion of persons at each ten-year date who had been in an occupation in a lower Social-Class ten years before. The samples reflect opposite patterns. For the lower, there has been a consistently rising proportion of persons entering this sample from below. This may in large part be explained by the greater numbers of working-class households in the twentieth century directories, but if this is so, it makes the pattern for the upper sample all the more surprising. For the latter, there seems to have been a marked fall in the proportion of soically mobile persons.

In the mid-nineteenth century, many, if not most, of the people who have here been described as the urban elite were recruited locally in a city like Leicester. Ellis (1948) has left a picture of nineteenth century Leicester which emphasises the closely knit, almost parochial nature of its society. Partly, this was encouraged by the strong hereditary links in such trades as hosiery, yet as the urban economy expanded after 1860, there grew up many more opportunities for local entrepreneurs in such trades as footwear and hosiery, where the structure of manufacturing, with specialist firms concentrating on particular stages in the process, encouraged small companies and an open and fluid industrial structure which continued into the twentieth century.

Though Leicester has spawned some industrial giants (like Corahs and Freeman Hardy & Willis in the nineteenth and the British Shoe Corporation in the twentieth centuries), the city remains characterised by the small enterprise and the family firm.

In developing a broad industrial base, Leicester has avoided too narrow an economic basis and too great a dependence on a single or a narrow range of industries and products. It is not dominated by any one great industry or giant firm. Of 1,000 members of the Chamber of Commerce, only 50 employ more than 500 people and 700 firms employ fewer than 100. It is balanced in the sense not only of being based on a wide variety of industries but also having within an industry a great variety of product and size of firm. (Pye, 1972, p. 13.)

It is possible to suggest that the fall in the numbers of socially mobile persons found in the upper sample after 1919 reflects a decline in local recruitment, which is paralleled by the switch from local mobility to losses from the directories, which actually represent an increased importance for inter-urban migration.[39]

Musgrove (1963) has illustrated the main reasons for this change. Now, not only do most of the professions recruit on a national basis (and there are more 'professional' occupations), but business and commerce have become more and more institutionalised as the scale of the economy has increased. Initially, the much greater numbers of young people going on to higher education have contributed to the process, because such an early departure from the home area often causes a break with the district of upbringing which is never re-established. Similarly, the growth of national career structures in both public and private organisations often demands frequent geographic mobility on promotion.[40]

Today, similar influences are operating in occupations in the lower sample. More and more jobs require qualifications and therefore are likely to need higher education. More and more, too, such jobs are being drawn into the national network of opportunities. Teaching is a good example, where local recruitment to many of the lower grades in the nineteenth century has given way to much wider horizons. Today, movement is very frequently looked on by teachers as a means of improving their salaries or status and the mobility of the profession has become an important factor in educational policy. The trend seems likely to increase. A much greater proportion of young people have been entering higher education in the last few years, and this alone would seem likely to foster a greater propensity to move between urban areas.

It is also an interesting corollary to ask whether the development of such avenues has been associated with the closing of some of the traditional paths of social mobility. The informal linkages between the upper-working and lower-middle classes were extremely important in a city like Leicester in the late nineteenth century. In the occupations selected in the lower sample, two good examples of such linkages can be recognised. One might be termed the clerk–manager route. Many of the managerial posts covered by the sample, e.g. in distributive and service trades, were recruited from clerks, salesmen, shop assistants and so forth. Another route was the craftsmen–master builder relationship. The character of the building industry has already been described as mainly consisting of small firms with little capital and even up to 1939, movement upwards (however temporary) was still easy.

One must stress that here only the middle-class section of the city's population has been examined in detail. At best, the occupations in the two samples are representative of the top 20 to perhaps 30 per cent of the population. Social mobility has been a factor in their development, especially in changing the migration patterns of the topmost families, but one must not exaggerate its role as an individual process. In the 1930s, when directory coverage was good, only about 15 per cent of the lower sample were apparently being recruited from lower social groups within the city.

Overall, in the period 1870 to 1940 Leicester grew enormously quickly and in terms of any assessment of changes in locational benefits, the city's spatial organisation altered rapidly and continuously. These changes accomplished and were accompanied by an equally continuous resorting of the urban population, and it is now necessary to tie together some of the factors involved by considering just what linkages were necessary on a year-to-year basis to effect such a relationship between the growth of the city, its social geography and the behaviour of individual households.

Short-term fluctuations in the rate of mobility

Fig. 4.19 illustrates the total change in turnover rates in the city between 1870 and 1950. The diagram was compiled on a similar basis to that for the turnover rates for each of the ten local sample areas, that is by taking every directory, drawing a 1 per cent sample of all entries, calculating how many persons were still at the same address in the next directory, deriving a turn-over rate from those who had moved, and then applying the compound interest formula to arrive at a rough per annum measure.

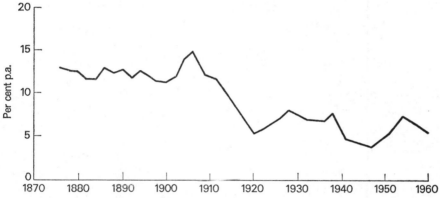

Fig. 4.19. Turnover rates in Leicester, 1870–1960

The obvious characteristic of the graph is that it reinforces the points already made about turnover in both the study of areas and the examination of the sample groups. In particular, one must stress the importance of the First World War in the city's mobility history. The average post-1919 rate of mobility is about 60 per cent of the rates common before 1914. The Second World War, on a smaller scale because of the lower base in 1939, produces a very similar effect. Both World Wars are periods when turnover appears to be at about 4 to 5 per cent per annum. It has already been suggested in 1870 that 4 per cent of all household changes occurred through natural events like death, and although it would be unwise to think that this might represent a constant, it makes one realise just how far the situation during the two wars must have represented an absolute minimum residential mobility.[41]

In this section, some sort of explanation will be sought not for the long-term decline in mobility but for the very obvious cyclical fluctuations in rates of movement. At this scale, the sort of socio-demographic changes already discussed in this chapter have to give way to a shorter-term explanation rooted in the everyday workings of the housing market.

The most simple hypothesis that suggests itself is that rates of residential mobility are directly related to the state of activity in the housebuilding market.[42] The greater the number of new houses being built in any period, the more new residential opportunities that will be set up and therefore more chains of movement will be initiated. Unless all new houses are occupied by immigrants or newly formed households, one would expect that building would create more 'elbow room' in the city. Fig. 4.20 illustrates the rate of

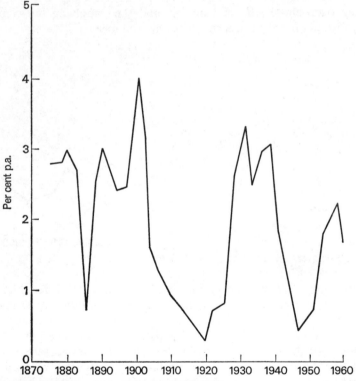

Fig. 4.20. New building rates in Leicester, 1870–1960

new building in Leicester measured by plotting the average annual percentage of the housing stock accounted for by new houses for all the periods covered by the directories between 1870 and 1950.[43] There have been violent and well-documented fluctuations in activity. Certainly, Leicester reflects, if not exactly parallels, the national pattern.[44] In the early 1870s, there was a high level of activity, followed by a slump in the 1880s, a recovery to a very high

116

peak in the years around the turn of the century, again followed by a recession which continued through the First World War. The effects of the war and the complete stop to private building have already been mentioned. The war was followed by another recovery which was interrupted by the economic crash of the period 1929–32. Housebuilding again recovered very strongly in the 1930s until the Second World War again imposed a complete halt. The war's end was again the signal for another boom in house construction.

Leicester and the national average have not been completely in phase. If anything, the swings seem to have been more violent at this local level. In Leicester, the peak of the early 1900s was much more significant than nationally, especially since it was followed by a quite catastrophic fall in building more marked than for the country as a whole. Undoubtedly, the severity of the slump in this period contributed to the particular strains that existed in the city's housing market during the First World War. After the war, Leicester saw the same pattern of peaks and troughs but relatively the 1920s were more important in the city than the 1930s, almost certainly because in the 1920s there was a boom in local authority building, which was concentrated within the Borough, whilst in the 1930s there was a boom in private housebuilding which was diffused over a larger area including such places as Braunstone, Birstall and Wigston.

In comparing the local rate of building to the rate of turnover, one must recognise that there are some similarities, but there are also differences. The two world wars represent the most obvious parallels, but that is not really surprising. One would expect periods of such exceptionally low activity in the building industry to be associated with very low rates of turnover. What is more significant perhaps are the areas of disagreement in the years before the First World War. The great building boom of the 1890s was not accompanied by any rise in mobility, yet after the turn of the century as building rates fell so the turnover rate actually rose quite steeply. This is initially surprising since it goes against the first, simple hypothesis presented.

Yet a little thought soon suggests that the relationship ought not to be so simple. If building represents the increase in the supply of dwellings, it does not follow that it accurately reflects the state of the housing market. Rather what is needed is that factor which will allow some measurement of the state of the current interaction between supply *and* demand. One possible measure exists in the number of empty houses in the city. It is known, for example, that in lieu of accurate information about the state of the housing market, small builders were very uncertain about the likely demand for their products. Saul (1962) has shown in the West Riding how they therefore tried to solve this problem by watching the rises and falls in the numbers of empty properties. The rate of empty houses is one of the few measures of the state of the market that can be obtained locally. Fortunately, Leicester does possess a sequence of statistics going back to the late nineteenth century which gives the number of dwellings defined as empty for rating purposes and does not therefore include merely those properties between owners. The proportion of such empty dwellings in the city ought to be a good measure of the degree of slack or shortage in the local housing market.[45]

Fig. 4.21 superimposes the rate of empty houses on the turnover rate already seen in Fig. 4.19. The degree of agreement is extraordinarily striking. As a simple way of comparing the two rates and also matching this to a comparison of the turnover rate with the rate of new construction, two Spearman Rank Correlation Coefficients were calculated, one between turnover and the rate of building; the other between turnover and the rate of empty houses.[46] The former produces a coefficient of $+0.061$, compared with $+0.767$ for the latter (which is significant at the 0.001 per cent level). The very good relationship between turnover and empty houses which the rank correlation coefficient produces is reinforced by the observation that the secular decline in mobility is paralleled by a secular decline in the average proportion of houses empty in the city at any one time. Just as it has been remarked that turnover and mobility do not seem to have ever been as high in the post-1919 period as in the years before, so one can see that the same is true of the proportion of empty houses in the city.

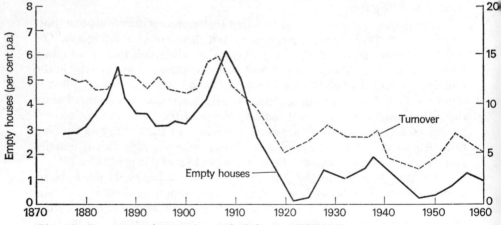

Fig. 4.21. Turnover and empty houses in Leicester, 1870–1960

In examining the detailed manner in which this relationship resolves itself, the most interesting period is that between 1890 and 1914, when the characteristics of the relationship are most highlighted. The 1890s saw a great boom in housebuilding in Leicester. There was a national upsurge in the industry, but it was exceptionally active in the city because Leicester was continuing to grow rapidly in the 1880s and early 1890s. In the four years between 1896 and 1900, there was no less than a 20 per cent increase in the total stock of dwellings in the city. Yet in looking at the rate of empty houses, it seems very likely that demand was able to continue to soak up all this

118

additional property right up till the turn of the century. The peak year for new construction was 1898–9, and the rate of empty houses only began to rise steadily from then on.

Building continued at a high rate into the early years of the new century, so if Leicester builders were using the rate of empty houses as a measure of demand, it would appear that they overstepped themselves. Certainly, from 1900 onwards, new construction would have appeared to have been feeding the total of empty houses in the city, and by 1907, there were no less than 3000 empty dwellings (about 6 per cent of the total). One obvious explanation would lie in the character of the housebuilding industry. Embarked on a particular phase of development, because of their financial commitments, builders in this period could not terminate their operations overnight. Housebuilding has never been an instantaneous process – the leads and lags in the industry caused by the necessary gap between starts and sales have always been a problem. Is one therefore to assume that the continued rate of new building after 1900 merely resulted in a series of catastrophic failures on the part of building firms? That assumption is far too simple. Some builders did go bankrupt, but one must realise that so far the discussion has centred on the aggregate level of supply and demand in the city.

The period immediately after the turn of the century represents the most extreme case of the manner in which building cycles and associated peaks and troughs in the city's turnover rate interact with the successive waves of social change. The mechanics of the situation are excellently illustrated thanks to the ever-valuable statistical information provided by the City's Medical Officer of Health. During this period before the First World War, he not only recorded the total number of empty houses in the city but also gave some details as to their location.

Fig. 4.22 shows the proportions of empty dwellings as a percentage of the total houses in each ward of the city in 1901 and 1909 respectively; that is at the point where the market seems to have been at peak demand and also eight years later when there was a large overall surplus of property in Leicester. There has been an across-the-board rise in the city, but empty dwellings are not concentrated on the periphery where new construction had been taking place, but rather in the central wards – Newton, St Margarets and Wyggeston – where the bulk of the housing was much older. Indeed, Aylestone where there had been concentrated much of the recent building in the last years of the nineteenth century, had actually experienced a decrease in the proportion of empty properties. The Medical Officer of Health was well aware of the disproportionate amount of empty housing in the central area, and he remarked of such dwellings that '74 per cent are cottages and a very large proportion of these are very old and of an inferior class' (MoH's Annual Report, 1909, p. 28). The consequences of this distribution in the light of the city's development system may be considered as part of a general model of social and spatial change in the city.

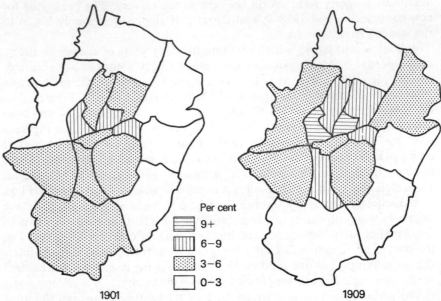

Per cent

▤ 9+

▥ 6–9

▦ 3–6

☐ 0–3

1901   1909

Fig. 4.22. The rate of empty houses by wards, 1901 and 1909

Such a model begins with the existence of the very well known cycles of activity in the building industry which have been much studied by economic historians (Saul, 1962, pp. 119–37; Habbakuk, 1962, pp. 198–230; Parry-Lewis, 1965). So far as the nineteenth century is concerned, these building cycles can be adduced to variations in home and foreign investment and the rate of emigration (Cairncross, 1953, pp. 27–33) as well as to increasingly important domestic factors such as the character of internal migration and a greater stability of finance for housebuilding which became more important in the period 1870–1914 and which had as one of their affects the lengthening of the cycles and produced a much more complex relationship with shorter-term business cycles (Habbakuk, 1962, pp. 228–30). At the detailed level, building cycles are likely to be strongly influenced by local rates of immigration which will distort these local cycles away from the national trend (Robson, 1973, p. 124), and this has been seen in Leicester with the relatively greater importance of the peak of activity in building in the early 1900s which reflects the high levels of immigration to Leicester in the 1890s.

Such cycles in building derived from variations in the general level of the economy and local growth of the urban population are imperfect in their detailed relationship to the level of demand, owing to the inevitable leads and lags in the housebuilding industry. As a consequence, the housing market varies considerably about an equilibrium, with tightness, reflected in a shortage of property, alternating with slackness, when there is a surplus. These up and down swings ought to be related to the interaction between housing costs and the average level of real incomes and this may be suggested as the main driving force behind variations in mobility in the city and the speed of spatial and social change.

The level of mobility in the aggregate fluctuates in direct correlation to the amount of slack in the housing market; being highest when the surplus of property is greatest and the cost of property relatively lowest. Furthermore, it has been seen that these fluctuations are associated (at least so far as the period before the First World War is evidence) with movements up the rungs of the housing market ladder which are reflected in a general movement which is outward and 'upward' to better quality property. It is likely that there must have been a third component to the cycle, which has not been examined here, which was a demolition/conversion cycle under which the vacated property was replaced (usually one must assume by the expansion of the city's commercial heart). General evidence for this has been presented as early as before 1859 with the growth of a central business district and the decline in central area residential properties, the continuing expansion of the central business area has been traced in the maps which illustrate the overall social geography of the city, and the more detailed picture of falls in central area residential properties can be repeated between 1891 and 1921. Evidence from the Census demonstrates that over this period which covers the cycle of the turn of the century years, the slump in housebuilding and the conditions of the First World War, there was a general drop of 40 per cent in the number of residential properties in the two wards of St Martins and Newton. The demolition of such properties probably thereby set up the conditions for the start of the next cycle by encouraging a shortage of property as the city continued to grow.

Of course, the evidence for the operation of this model is fragmentary over a long period of time and there are complications. The ten sample areas, for example, suggest that whatever the overall correlation between mobility, building and the state of the market, there must have been local variations. This is sensible; any process operating through time like this must have incurred leads and lags which are so familiar in other parts of the market, and it seems likely that the variation in levels of mobility seen in the ten areas is representative of the way in which the cycle's effects would have spread over the city, successively from newer to older areas, with the consequent peak of mobility being somewhere between the peak of building and the trough of the maximum number of empty properties. Similarly, the upper sample areas will represent districts in which the relationship was likely to be less well-developed for all the reasons already put forward for the spatial and housing distinctiveness of such areas.

But in general, new building was related to two main sources of demand – the growth of the urban population and rising real incomes. The latter had to occur to translate a demand for a higher quality of dwelling in new property which might come from some other source such as the public health lobby, into bricks and mortar. Indeed, after 1860, new property was increasingly aimed at families higher and higher up the social scale. At best, it has been seen how the cheaper properties that were built between 1870 and 1914 were almost exclusively aimed at a mixed lower-middle-class/artisan population who maintained a constant relationship with such new housing on the periphery, which implies a high rate of spatial and social change for these groups and all below them in the social scale and the housing market.

There is little evidence, however, that such social change proceeded in anything like the manner suggested by the invasion–succession model. Rather, it was a continuous process, which operated at different speeds controlled by the housing-market/mobility model just outlined. Dyos & Reeder (1973, p. 369) are quite right to stress that suburbs 'were above all the strategic component in the housing of the urban community', and one cannot but be aware of the way in which virtually all the elements in the housing market (with the possible exception of the very high-value dwellings) were linked in a process which clearly subordinates any manipulation of the market by individuals or groups (except again those right at the top) to the need to maintain the general equilibrium of spatial and housing costs in the face of very rapid urban growth.

Perhaps the most significant problem associated with the simple model of building and mobility outlined above is its applicability after the First World War. It has been shown that the relationship between mobility and the rate of empty houses continues after 1920, but that relationship appears much less useful in terms of any pattern of social and spatial change. Rather, the inter-war period appears characterised by a generally much lower level of mobility which may have been especially associated with a general shortage of property, created by the First World War, but not remedied in the post-war period. It is likely that the whole pattern of changes in the provision and operations of the housing market described at the beginning of this chapter, which have been cited as causes for the disappearance of long swings in building activity (Aldcroft & Richardson, 1968;1969) were similarly responsible for the collapse of any relationship between the workings of the market and mobility.

It is clear that Leicester has undergone great secular changes in the role of mobility and movement in the city and their relationship to both the housing system and the social geography. But before any attempt is made to account for the consequences of such changes, a more detailed analysis of the processes at work in the contemporary city is needed, and this will be done by looking at the growth of the city since 1945 and then by investigating the detailed pattern of movement for a single year, as was done for 1870.

CHAPTER 5

HOUSING AND MOBILITY IN POST-WAR LEICESTER

The post-war housing market

The Second World War created a situation similar to the First. The supply of labour and materials was again limited by strategic considerations to the construction of houses for war-workers. The normal processes of house-building came to a complete halt and the maintenance of existing properties suffered with an inevitable deterioration in their condition. In addition, there was the damage caused by air-raids. Nationally, a quarter of a million houses were destroyed, another quarter were rendered uninhabitable, and perhaps three million damaged in some way (Cullingworth, 1966, p. 29). Once again, the war brought with it the introduction of a rent and mortgage freeze.[1] Cullingworth has suggested that in 1939, there was a rough balance nationally between the supply of and demand for accommodation, but by 1945, it could be estimated that there was a shortage of three-quarters of a million dwellings.[2]

The Labour Government elected in 1945 concentrated on the construction of properties to rent by local authorities,

It was a tenet of a strong Labour administration that the government should not only control and plan house production, but also ensure that the allocation of dwellings was on the basis of the most urgent need, and not of capacity to pay. Local authorities were to be the instruments for house production and allocation. (Smith, 1971, p. 4.)

The 1949 Housing Act removed the obligation on local authorities to provide dwellings only for the 'working classes',[3] and thus implied that they were to be responsible for the total housing needs of their areas. Overall, local authority powers were greatly strengthened by the 1947 Town and Country Planning Act which gave them the ultimate responsibility over land allocations,

the 1947 Act contained some of the most drastic and far-reaching provisions ever enacted affecting the ownership of land and the liberty of an owner to develop and use his land as he thinks fit. Indeed, after July 1 1948, ownership of land, generally speaking, carries with it nothing more than the bare right to go on using it for its existing purposes. The owner has no right to develop it; that is to say, he has no right to build upon it and no right even to change its use. (Heap, 1973, p. 12.)

The statutory planning system has been much amended since 1947 (notably in 1968 and 1972) but its original intentions towards the public control of land use remain.[4] Of necessity, the planning system has been designed with a negative rather than positive slant and there have been shifts in the relative balance between authorities and developers. But the powers of local planning authorities to draw up development plans to guide the patterns of growth in their areas through the distribution of land uses and more especially their ability to enforce these plans through a statutory system of development control, have necessarily formed a very important element in the recent growth of British cities.[5]

Private building did not recommence on any scale after the war until the return of a Conservative Government in 1951, and their removal of the system of building licences in 1954. Since 1954, all governments, with some differences in political emphasis, have subscribed to a dual system of development which has built council houses to let and owner-occupied houses for sale, side by side. Of course, this was the basic system that had existed in the inter-war period, and it is most convenient to look at the post-1945 housing marker within a tenurial framework (Cullingworth, 1966; Donnison, 1967; Smith, 1971; 1974).

Owner-occupation

The recommencement of private building soon led to a rapid recovery in the buoyancy of this sector, and by 1960, private builders were as active as local authorities, and since that date, they have generally exceeded the efforts of the public sector, reaching a peak of nearly a quarter of a million new houses a year in the late 1960s.[6] As before 1939, financing of new construction by private enterprise has mainly been through the raising of mortgages by potential buyers. The rate at which first time buyers have been able to enter the market has become an increasingly important factor in generating new construction. Building societies remain the most important source of mortgage finance, providing 85 per cent of all loans advanced for house purchase in 1971, but the role of local authorities and insurance companies has been significant.[7] The availability of mortgages has varied with the state of the national economy and has so constrained the effective demands from potential owner-occupiers in periods of tight credit that the cycle of booms and slumps in building has continued.

The building industry itself has changed little. The majority of firms remain small and their contribution, individually, to the total product minimal. Methods remain traditional, despite efforts to introduce industrialised system-building techniques. Some of the larger construction companies have diversified into housebuilding in the post-war period and this has encouraged the development of economies of scale which can be recognised in the standard designs that such firms have applied on estates all over the country.

In a time when inflation has been a consistent feature of the economy, house purchase has been often the cheapest form of accommodation, despite a secular trend for interest rates to rise, because the costs of servicing an initial purchase loan have fallen steadily as opposed to the rising capital value of one's property. But entry to owner-occupation has been restricted by income and the ability to obtain a mortgage has remained outside the creditworthiness of many.[8] Nevertheless, the growth in owner-occupation has been striking in the post-war period. In 1947, about 27 per cent of the national housing stock was in this tenure, some 3.9 million dwellings: by 1971, half of all the properties in the country were in owner-occupation, about 9.5 million dwellings. About 70 per cent of this increase has come from new construction, but between one and a half and two million properties have passed into owner-occupation from other tenures.

Local authority housing

Building by local authorities has been the other main means of adding to the stock. For the greater part of the post-war period, financing of local authority building schemes came from the rates and subsidies by central government provided through sixty-year, fixed-interest Treasury Loans.[9] Costs and standards have been controlled by the central authority. Costs have been assessed through the 'yardstick', a complicated formula designed to avoid any open-ended committment on the part of government, such as occurred under the 1919 Housing Act.[10] The standards of the inter-war period, laid down by the Tudor Walters Report, have been twice up-dated. In 1944, the Dudley Committee on the Design of Dwellings improved standards both of size and facilities and their recommendations formed the basis for local authority houses till the 1960s, when they were superseded by those of the Parker Morris committee.[11]

Local authorities' activities have concentrated on the same two problems that they faced in the inter-war period – an unmet demand from those unable to meet the costs current on the open market, and slum clearance. Surveys at the time of the 1954 Housing Act suggested that there might be one million unfit dwellings in Britain,[12] but in practice, rising standards and poor levels of upkeep in the cheaper sections of the market have led to the problem continuing to run ahead of efforts to solve it. In general, in the 1950s and early 1960s, local authorities took a great part in replacing cleared slum property with new council dwellings, especially in high-density/high-rise flats. Private enterprise was discouraged from taking any part in this process by high costs and a lack of demand and the main competitors for such sites have been commercial property developers (NEDC, 1968). In the late 1960s, such comprehensive redevelopment began to fall into disfavour in some quarters because of high costs, low returns and the social problems which it was said had been produced in the new estates. Since the 1969 Housing Act, much greater emphasis has been put upon the improvement of older properties through the wider provision of grants by central government and the local authorities' powers to declare General Improvement Areas.[13]

Unfurnished private renting

With a few minor exceptions, new building to let by private landlords has been virtually non-existent since 1945. Exceptions have been a small amount of building for a luxury market, especially of flats in Central London; building for certain special cases like servicemen and forestry workers; and the activities of housing associations. Housing associations are the heirs of the Victorian philanthropic housing trusts and have received considerable government assistance in recent years. They have been encouraged to provide 'cost-price' housing for special groups, and have been called 'housing's third arm'. In practice, they remain a very small element in the market and do not account for more than between 1 and 2 per cent of the total stock.[14]

The continuing lack of activity in this rented sector of the market can be explained by much the same reasons that accounted for its decline in the inter-war period. The tax system has continued to operate against the private landlord and just as for much of the period there was a movement away from interest-earning stocks to capital accumulating equities, so there has been a shift from letting to owner-occupation. Rent controls have continued to be a much discussed factor. Not until 1954 was there any attempt to alter the wartime restrictions when there was some allowance made for rent rises to meet deterioration in the quality of many of the cheaper properties in the market. The 1954 Act only touched the problem, but in 1957 a serious effort was made to deal with privately rented dwellings. The 1957 Rent Act was an extremely controversial measure, which really seems to have had very little influence on the long-term evolution of the market. It was very similar to the 1923 and 1933 Acts in allowing for progressive decontrol. Higher rated properties went out of control immediately; vacant possession of all properties brought with it decontrol; and the rents of cheaper properties were raised. The ambitions of the Act were once more to inject a new incentive for private landlords to come back into the market, to encourage greater mobility and provide a better match between households and dwelling size. Outside London, the Act may have had some effect in improving the upkeep of higher-rented propertes, but did little to prevent or slow down the continuing movement away from this tenure as landlords sold up on gaining vacant possession (Cullingworth, 1963). In London itself, the Act came under very heavy fire for supposedly encouraging malpractices in the private rented sector. The Milner Holland Report of 1965 was instrumental in the introducton of the 1965 Rent Act which gave tenants security of tenure and set up a national system of tribunals to assess the 'fair rents' for such properties.[15]

Between 1953 and 1969, nationally, the number of privately rented dwellings fell from 7.2 to 2.9 millions. Immediately after the Second World War, this tenure had accounted for 55 per cent of all properties, by 1971, it was down to 15 per cent. Many of the cheapest properties had been cleared by slum clearance,[16] but equally important has been the switch produced by the sale of dwellings to sitting tenants or on vacant possession. In 1971, it is known that 55 per cent of all the properties built before 1919 (the vast majority of which were originally let) were in owner-occupation (GHS, 1973,

126

p. 98). In addition to this switch, there has been a substantial movement into the alternative of furnished letting.

Furnished letting

The growth of furnished letting has been a very important factor in central city housing markets in the last thirty years. For the greater part of this period, furnished letting possessed legal and economic advantages for landlords which encouraged its growth. It is still only a small element in the national housing stock, perhaps 3 to 4 per cent of the total. Where it is important is right in the centre of cities, where it now often provides the main source of accessible, cheap accommodation. 60 per cent of all furnished tenancies in 1971 were in London, but all major cities have seen the growth of this tenure since 1945. Attitudes to the emergence of this tenure have varied. Many have blamed it for exacerbating the poor housing conditions which are so concentrated in such central areas. Others have thought it an unfortunate necessity. The Francis Committee on the Rent Acts took this attitude,

without doubt these are the areas where most of the small minority of greedy and unscrupulous landlords thrive. But it must be borne in mind that even these people provide a home for many tenants – often not very good tenants – who would otherwise be homeless. No doubt, it is often an inferior and overcrowded home, but it is better than no home, and preferable for a tenant with a family to living indefinitely in welfare accommodation. (Francis Report, 1971, p. 221.)

Circumstances have again changed recently and the 1974 Rent Act has substantially eliminated the difference between furnished and unfurnished tenancies.[17]

Leicester in the mid-1960s

In the period since 1945, Leicester's position as a centre of rapid growth has continued. The Second World War brought a great upsurge in engineering industries, often of a highly technical and specialised nature, and these have been contributory to the city's state of economic health in the last thirty years;

the Leicester region has experienced marked industrial prosperity and growth in the post-war period. The already diversified industrial base of the late 1940s has been, if anything, further widened during the last quarter of a century. (Mountfield, Fagg & Gudgin, 1972, p. 403.)

Three sectors of manufacturing remain important – hosiery with about 15 per cent of total employment; footwear with about 7 per cent, and general engineering with another 15 per cent. Services have also grown, both because of the city's continuing importance as a regional centre, but also in connection with such special functions as the University.

127

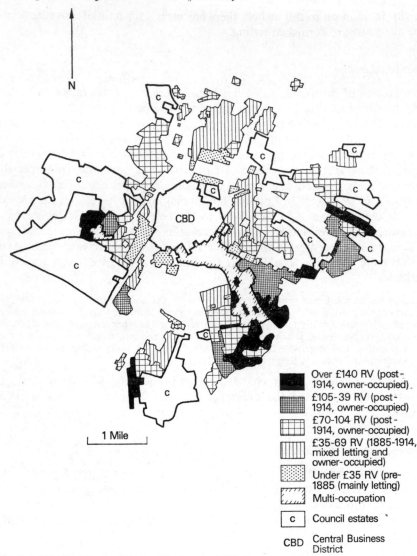

Over £140 RV (post-1914, owner-occupied)

£105-39 RV (post-1914, owner-occupied)

£70-104 RV (post-1914, owner-occupied)

£35-69 RV (1885-1914, mixed letting and owner-occupied)

Under £35 RV (pre-1885 (mainly letting)

Multi-occupation

c   Council estates

CBD   Central Business District

Fig. 5.1. The housing structure of Leicester, 1963–4

Fig. 5.1 shows the area of the city in the middle 1960s, with the main elements of its housing structure outlined.[18] The map was compiled from a variety of sources, the main one being the 1963 Rating Valuation Rolls which were used to assess the relative value of private property in the city. Individual street blocks were allocated to five value categories on the basis of their median value. The tenure of areas was derived from two sources. Local authority properties are very easily identified from the Rating Valuations where they are listed separately, and for private property, the dominant tenure was estimated from the enumeration district material of the 1961 and 1966 Censuses, which were also used to identify areas of multi-occupancy in the city.

The much more complex geography of housing in the post-1945 period is clearly illustrated. The major feature in the city is the physical scale of the new local authority housing estates. Between 1945 and 1966, the local authority built over 16,000 new dwellings as part of a scheme to replace the estimated 18,000 sub-standard properties in the city.[19] 8000 houses were demolished in clearance schemes, most of them around the central area, especially on the northern and eastern sides where much of the great mass of property built up to 1875 has been removed.[20] As well as central area redevelopment schemes, there has been a vast amount of building on the periphery, and an almost complete ring of estates now surround the city.

These estates mark the carrying out of the land allocations of the late 1940s and have effectively blocked off the avenues of expansion that had been followed by private builders in the 1930s. Consequently, only 30 per cent of all building within the city boundary since 1945 has been by private enterprise. Only in the south-east has such development been more important, and much has taken the form of infilling the looser patterns of the growth of the inter-war period in areas like Belgrave and Evington. Because their activities have been restricted within the city, private builders have mainly looked outside its boundaries for land. Though the City Development Plan started off with a fairly restrictive set of assumptions – 'Broadly, the planning objective is to secure a satisfactorily terminated and rounded off urban area based on the present city, which would be interspersed with green wedges and surrounded by an adequate green belt where possible' (City Development Plan, Written Analysis, 1952, p. 68) – these have not been borne out by the very rapid rate of growth in the city's economy and population. Between 1945 and 1966, the population of the Leicester Sub-Region rose by some 70,000, even though the population of its central city, Leicester County Borough, remained stable. Most of the development has occurred just over the city boundary, especially in such settlements as Oadby, Wigston, Thurnby and Birstall. Oadby and Wigston, for example, have seen as much private building together as Leicester itself. Oadby has tripled in size and Wigston doubled since 1945, and neither place is now physically distinguishable from the main mass of the city. In recent years, the activities of private builders have spread further afield, as increasing car-ownership and cheaper sites encouraged development of numbers of speculative estates in a ring of villages five to ten miles from the edge of the city. Anstey, Kirby Muxloe, Desford, Blaby and Countesthorpe are all examples of this trend. Even further out, villages and small towns have been influenced by growing numbers of Leicester commuters seeking homes. In all, the three rural districts which existed before April 1974 and which surrounded the city, have seen 50 per cent more private building since 1945 than Leicester, Oadby and Wigston combined.

Owner-occupation has also spread back into the inner city areas through that process already mentioned at the national scale, whereby landlords have sold off older houses formerly let. It is impossible to specify in detail the scale of this process since 1939 in Leicester, since exact figures as to the tenurial composition of the stock before the Second World War are not available, but a rough calculation on the basis of slum clearance activities and the results of the 1971 Census suggests that between 25,000 and 30,000 unfurnished tenancies must have been lost. Between a third and a half would have disappeared because of demolition, but the rest were lost through conversions, predominantly to owner-occupation, but in some cases to furnished lettings.

It is not surprising that the Enumeration District data for 1961 and 1966 should have indicated a strong association of the remaining unfurnished tenancies with the very oldest and poorest quality housing in the city. They are especially to be found in those areas where the local authority might be expected to mount a clearance scheme in the near future. In almost all areas of the city built after 1885–90, owner-occupation has become the majority tenure, and in all, unfurnished tenancies had fallen to only about 18 per cent of all properties by 1971. Indeed, if present trends continue, the combined affects of slum clearance and the withdrawal of landlords seem likely to make this tenure virtually extinct by the early 1980s.[21]

The special role of furnished tenancies must be mentioned. In 1971, they only accounted for about 7 per cent of all the households in Leicester, but in comparison with the unfurnished sector, this tenure had been expanding rapidly in the 1960s. Though there are no detailed figures, the pattern of houses broken up into multi-occupancy (with which this tenure is especially associated) suggests that it could not have accounted for even one-hundredth of the city's dwelling stock in 1939, and was probably not much more significant even so far as the let sector of the private housing market was concerned. By 1971, such furnished tenancies accounted for a third of all private lettings. The present association with multi-occupancy is a strong one. Half of all the households in Leicester in 1971 who shared accommodation were in this tenure. A third of all furnished tenants were in such shared properties,[22] and half of them did not have the exclusive use of a bath or WC.

The inter-war concentration of such households in the inner, south-eastern sector of the city has intensified since 1945, but has also spread out. Whereas before 1939, such multi-occupation was generally confined to the larger properties built before 1880 and to the areas north-west of the Midland Railway, it has now spread out into such areas as Stoneygate and Clarendon Park, and also to subsidiary centres in the west of the city. Three trends seem to have lain behind the growth of this tenure in the last thirty years. The first has been the general shortage of cheaper property and the restrictions on entry to the dominant tenures of owner-occupation and council letting imposed by income or policy. The second has been its relative attractions for much of the period to landlords, in preference for the much more severely constrained unfurnished letting sector. Thirdly, there have been demands from such special groups as students and Commonwealth immigrants who have sought cheap property and have been prepared to put up with such inconveniences as sharing.[23]

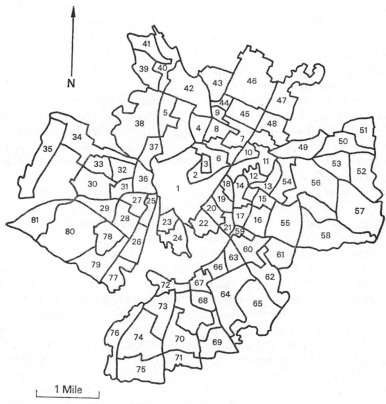

Fig. 5.2. Sample areas for the 1963–4 analysis

In sketching out the spatial organisation of the city's housing market, it is obvious that a far greater complexity has been displayed in the post-1945 period, due mainly to the proliferation of legislative and administrative interventions in the market, the growing importance of tenure as a feature distinguishing particular areas of the city, and the breaking down of the relatively simple spatial patterns that characterised the city before 1914. Before considering the patterns of mobility and migration that operate in this situation, it would be useful to construct a more explicit framework within which such behaviour might be examined. This is especially true in terms of the relationship of the social and demographic structure of the city with the pattern of housing, and to develop this context the very familiar technique of a principal components analysis of Census Enumeration District data will be used.

It is not proposed to go into any great detail here as to this well-known technique.[24] There has been considerable discussion about its mathematical structure and validity especially in terms of either generating hypotheses about urban society or of producing 'objective' dimensions of urban spatial structure.[25] Here, it is used more simply to provide a structured framework for the social geography of the city in which human and housing factors are linked to form the context for the central interest in mobility and migration. It is therefore used in a way much closer to the concern with regionalisation problems than with studies of factorial ecology.

At the time when the analyses were carried out, the only small area information available was from the 1961 and 1966 Censuses.[26] The 1961 Census was carried out on a 100 per cent base so far as the demographic material was concerned, but is only a 10 per cent sample for the so-called socio-economic information. It is therefore marginally more reliable than the totally 10 per cent, 1966 Sample Census. But the latter does contain some more up-to-date information, especially with regard to the spread of recent Commonwealth immigrants in the city, which was going on particularly rapidly in the early 1960s. It was therefore decided to use information from both Censuses.[27] It is fortunate that although the 1961 Enumeration Districts were smaller than those in 1966, and there was a boundary change in the city between the two dates, the two sets of areas used can be fitted together. Because of the general constraint imposed by the need to be able to use these districts later in the analysis of migration, they had to be large enough to produce a sufficient sample of one-year's movers, and fig. 5.2 shows the eighty-one areas that were constructed to form the areal basis for both the principal components and migration studies. They contain between 800 and 1500 households and though their gross area varies, the amount of residential land in each is of the same scale. It is difficult to comment on their homogeneity. Efforts were made to introduce some sort of rationale in terms of their composition of housing tenures and types and the eighty-one areas are a necessary compromise. Census districts take some account of major morphological barriers in the city (if only to ease the task of the enumerators) and such features as the boundaries of the larger council estates are often reflected in their shape. But especially in the older areas of the city, either such homogeneity does not exist or its scale cannot be reproduced when the need to make the areas large enough for the migration work has to be borne in mind. Table 5.1 lists the thirty-two variables used in the principal components analysis; fifteen deal with housing, the rest with socio-demographic elements especially relevant to the life-cycle career-mobility model.

TABLE 5.1   *Principal components analysis, 1961/1966 data: variable list*

1. % owner-occupied households
2. % dwellings rented from the local authority
3. % households rented unfurnished
4. % households rented furnished
5. % dwellings with less than four rooms
6. % dwellings with more than six rooms
7. Mean persons per room
8. % households living at less than 0·5 persons per room
9. % households with exclusive use of all four household arrangements
10. % dwellings in multiple form
11. % households with less than four rooms
12. % dwellings with a garage (1966)
13. % households sharing
14. % dwellings unoccupied
15. % households with more than six rooms
16. % adult males in Social-Classes I and II
17. % adult males in Social-Classes IV and V
18. % households with a car (1966)
19. % persons aged under 5
20. % persons aged 6–19
21. % persons aged 20–9
22. % persons aged 30–44
23. % persons aged 45–59
24. % persons aged 60+
25. % households with no family
26. % households with two families
27. % households with one or more persons of pensionable age
28. % of females aged 25–64, economically active
29. % households with more than six persons
30. % persons born in 'New Commonwealth' countries (1966)
31. % persons born in Ireland (1966)
32. % households with two persons or less

The principal components analysis

For each of the components which the analysis produced, two elements are highlighted – the loadings which identify the variables most closely connected with that component, and the scores which identify the areas.[28]

Component 1

+0.8548   24. Persons aged over 60
+0.7949   8. Households living at less than 0.5 persons per room
+0.7348   27. Households with one or more persons of pensionable age
+0.6794   32. Households with two persons or less
+0.5096   25. Households with no family
−0.5205   7. Mean persons per room
−0.5833   20. Persons aged 6–19
−0.7216   22. Persons aged 30–44
            Total variance 16.78%

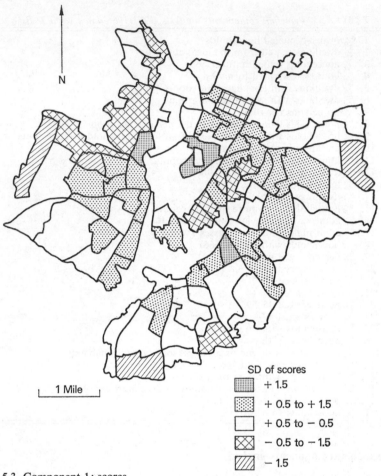

N

SD of scores
- ▦ + 1.5
- ▨ + 0.5 to + 1.5
- ☐ + 0.5 to − 0.5
- ▧ − 0.5 to − 1.5
- ▨ − 1.5

1 Mile

Fig. 5.3. Component 1: scores

It is an important preliminary point to remember that all components are best seen as dimensions or spectra, stretched across the width of the data. Component 1 identifies a clear distinction between areas associated with small households and old people and those with larger households and younger families, which is shown in fig. 5.3.[29] The spatial distribution of this component is a centre–periphery contrast. No post-1919 area loads in a significantly positive manner on it, and the highest negative scores are to be found in such new areas as the modern council estates or the post-war private housing in South Knighton (Area 69). The highest positive scores occur in a ring around the centre of late-nineteenth-century terraced housing areas, and the highest scores of all were actually in districts (Areas 6 & 37) undergoing redevelopment in this period. There are two exceptions to the pattern. The

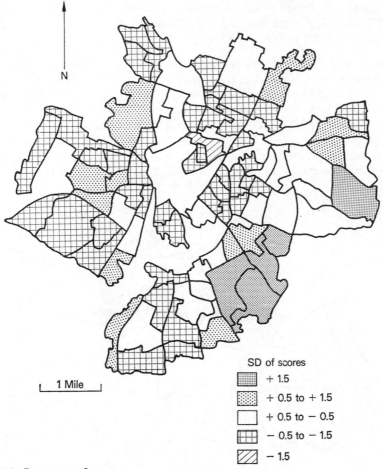

Fig. 5.4. Component 2: scores

districts which have already been seen as the heart of the multi-occupied, furnished letting sector of the housing market do not appear with positive scores and neither do a block of areas in the Spinney Hills district (Areas 12, 14 & 18).

Component 2

+0.9039	12.	Dwellings with a garage
+0.9008	16.	Adult males in Social-Classes I & II
+0.8500	18.	Households with a car
+0.7583	1.	Owner-occupied households
+0.5531	15.	Households with more than six rooms
+0.5481	9.	Households with exclusive use of all four household arrangements (bath, WC, hot and cold water supply)

—0.6613   28. Economically active females aged 25–64
—0.8901   17. Adult males in Social-Classes IV & V
Total variance 14.87%

Component 2 is very obviously related to social class, standard of housing (and presumably income). The pattern of scores shown in fig. 5.4 illustrates a pattern which the earlier chapters have made very familiar. Very high positive scores occur in Stoneygate, Knighton and in the newer parts of Evington. More moderate positive scores are associated with the great mass of inter-war, private housing (e.g. Humberstone, Evington and parts of Westcotes). Every positively-scoring area which appears as significant was built after 1919 (with the exception of Area 16, which straddles the date boundary).

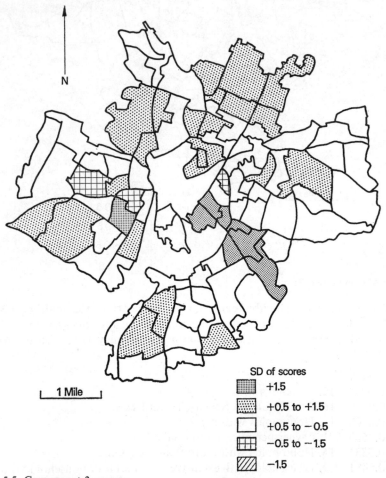

SD of scores
+1.5
+0.5 to +1.5
+0.5 to −0.5
−0.5 to −1.5
−1.5

1 Mile

Fig. 5.5. Component 3: scores

Naturally, in view of these positive scores, the lowest negative scores are in the poorest areas with the poorest quality housing in the early 1960s (e.g. the remnants of the pre-1870 areas that had not yet been cleared on the northern and eastern edges of the city centre).[30]

The first two components account for nearly a third of the total variance in the matrix of thirty-two variables, and it is especially interesting to contrast their apparent roles with regard to housing. Component 1 appears to be aligned transverse to the structure of the market, especially with regard to tenure. Only one tenure, unfurnished letting, loads even moderately significantly on this component probably because it is concentrated in the inner, pre-1919 areas. By contrast, Component 2 has a strong housing element with owner-occupation appearing as a significantly positive variable and the other three all having negative scores. This certainly argues that housing has a stronger association with social class than with demographic factors.

Component 3

+0.9519	13. Households sharing
+0.9297	4. Households renting furnished
+0.7626	25. Households with no family
+0.6668	11. Households with less than four rooms
+0.5043	31. Persons born in Ireland
	Total variance 10.99%

This component is obviously very strongly determined by the city's housing structure being those areas (illustrated by fig. 5.5) associated with furnished letting and multi-occupancy. But it is also a component with a strong demographic element. It is especially correlated with those households which are not formed by families (i.e. they are not based on a married couple) and which are composed of such groups as students and young people sharing. This role as an area for a transient population is reinforced by the importance of the Irish population in such districts, and the dominance of people aged 20 to 29 years old. The pattern of scores for this component shows very well how though it is still centred in the older area of Victorian town houses, north-west of the railway, such multi-occupation is spreading out into such areas as Stoneygate, and also into subsidiary concentrations like those associated with the larger properties fronting main roads in Westcotes and Spinney Hills.

Component 4

+0.9303	5. Dwellings with less than four rooms
+0.8062	10. Dwellings in multiple form
+0.5380	11. Households with less than four rooms
−0.7834	6. Dwellings with more than six rooms
	Total variance 10.45%

137

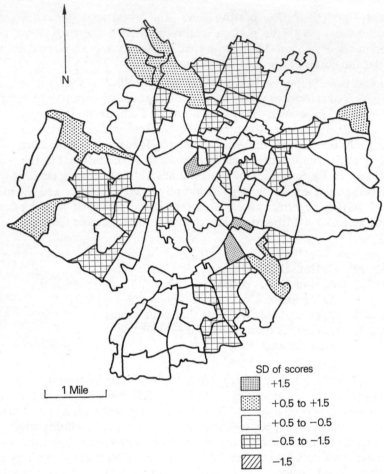

N

1 Mile

SD of scores

+1.5

+0.5 to +1.5

+0.5 to −0.5

−0.5 to −1.5

−1.5

Fig. 5.6. Component 4: scores

Component 4 (illustrated in fig. 5.6) picks out an element in the city's housing structure that has only become of importance since 1945. These are purpose-built flats (i.e. they are small dwellings in multiple structures not merely small households sharing larger dwellings). Two types of area appear on this component – council estates with high-density accommodation and private areas where some redevelopment has taken place and blocks of three or four storey flats have replaced older, low-density Victorian dwellings. This latter feature being especially important in parts of Stoneygate (e.g. Area 63).

Component 5

  −0.8096   29. Households with more than six persons
  −0.8103   26. Households with two families
               Total variance 8.33%

138

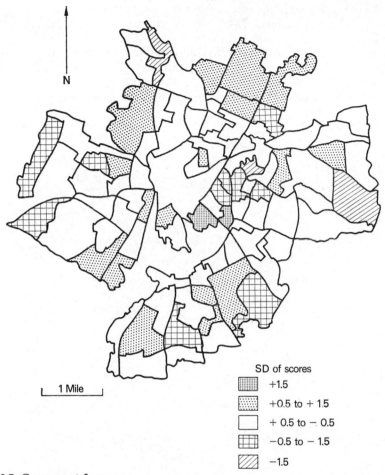

SD of scores

▓	+1.5
▒	+0.5 to + 1.5
☐	+ 0.5 to − 0.5
▦	−0.5 to − 1.5
▨	−1.5

Fig. 5.7. Component 5: scores

As one proceeds down the ranking of the components, not only does the degree of variance 'explained' fall, but the generality of the component shrinks. Component 5 (fig. 5.7) demonstrates this by picking out a dimension in the city associated with small families and a low density of occupation on the positive scores, and overcrowding and sharing on the negative. The latter types of area are concentrated in the poorer areas nearer the city centre, especially the Spinney Hill district, but also more surprisingly in certain of the council estates.

Component 6

  +0.6167   30. Persons born in 'New Commonwealth' countries
  −0.5053   20. Persons aged 6–19

−0.6720   2. Dwellings rented from the local authority
                Total variance 7.48 %

This component (illustrated by fig. 5.8) is one associated with the recent
settlement in Leicester of immigrants from 'New Commonwealth' countries,
and especially of Asians from India, Pakistan, Bangla Desh, and more
indirectly from East Africa. These groups became an increasingly important

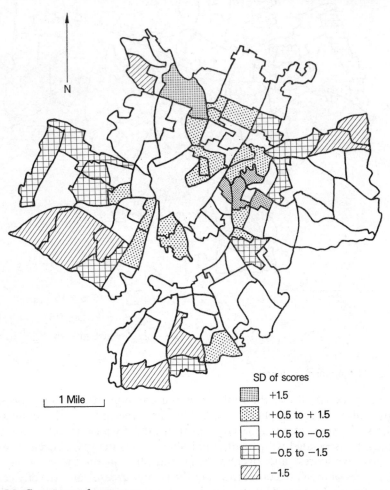

Fig. 5.8. Component 6: scores

element in the city's social structure throughout the late 1950s and early
1960s. Leicester was never a cosmopolitan city. In the mid-nineteenth century,
its Irish community never exceeded 2 per cent of the population and in 1951,
less than 4 per cent of the city's population had been born outside the United

Kingdom, yet by 1971, this proportion had risen to 12 per cent. In the interim, about 20,000 Asians had come to the city.

The first influx was especially dominated by large numbers of single men, who had come to work in the hosiery industry, but after the passing of the Commonwealth Immigration Act in 1962, a much higher proportion of families came to the city. The initial areas of settlement were very much in the multi-occupied/furnished-letting sector, but later they have spread out into

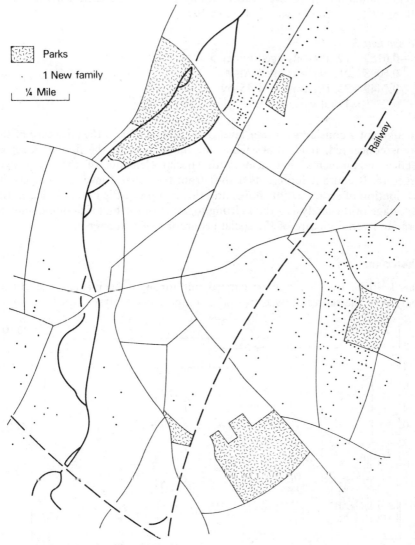

Fig. 5.9. New Asian families, 1963–4

the late-Victorian terraces of South Belgrave and Spinney Hills. Fig. 5.9 shows those houses occupied by Asian families in the year 1963–4.[31] In that year, no Asian family was situated outside this central area illustrated by the map, and this is reflected by the inverse correlation of immigrant households with local authority property, which is predominantly peripheral in location. Both owner-occupation and unfurnished letting have moderate positive correlations with this Component, and this ties in with the observed fact that Asian immigrants have been more willing to purchase their own property (Rose, 1969, p. 137; Davies & Taylor, 1970).

Component 7
+0.6153   19. Persons aged under 5
+0.5088   21. Persons aged 20–9
−0.7449   23. Persons aged 45–59
          Total variance 5.43%

Component 7 considers a lower level of analysis so far as the structure of the city is concerned. It identifies what one might take as the first of a set of detailed components that deals with special elements of the city's age-structure. Because it represents a shift from the general level of analysis, its distribution of scores is not shown and this component is not included in the next stage of the analysis – the knitting together of the first six components to form an overall picture of the social geography of Leicester.

Cluster analysis

This knitting together will be carried out through the means of a cluster analysis of the scores derived from the components analysis. Cluster analysis

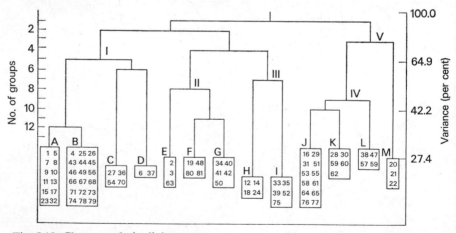

Fig. 5.10. Cluster analysis: linkage tree

is a technique of classification by which a matrix of similarities, in this case the mean Euclidean distances between the six sets of scores, is used as a basis for a hierarchical grouping procedure (King, 1969, pp. 198–204). Here, the Ward algorithm has been used to construct the linkage tree whose upper stages are shown in fig. 5.10. The Ward function minimises the within-group sums of squares at each step to decide the sequence in which the individual elements are to be fused.[32] This was not a problem which required a contiguity constraint, and fig. 5.10 also shows the proportion of total variance 'explained' by the successive within-group functions. As King points out, there is little analytic guidance for selecting those points in the linkage which are the most significant. Here, on an entirely empirical basis, two main levels have been identified as being of importance – one at thirteen groups, the other, higher level, at five groups. Fig. 5.11 maps the distribution of the eighty-one areas

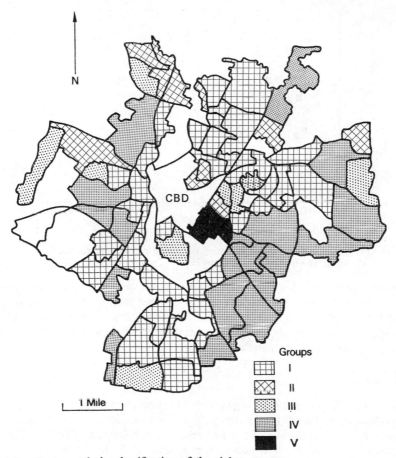

Fig. 5.11. Cluster analysis: classification of the eighty-one areas

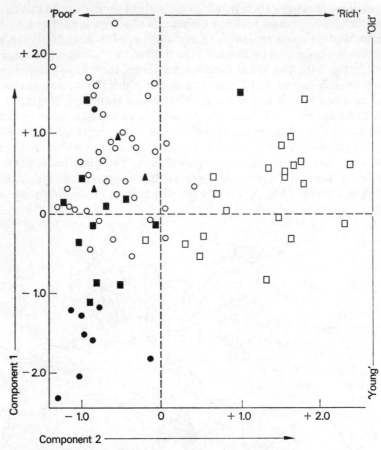

Fig. 5.12. Cluster analysis: classification groups and component scores.
Key: ○ Group I. ▣ Group II. ● Group III. □ Group IV. ▲ Group V

between these five higher-level groups, whilst fig. 5.12 plots the relative positions of the areas by their five groups against their scores on Components 1 and 2, which account for a third of the total variance in the original data matrix, and which might very approximately be associated with the life-cycle and career-mobility elements in the city.

Group I

Group I is the largest of the higher-level groupings in terms of the number of areas included. Two-fifths of the enumeration districts in the city are in this group. It was also the last of the five groups to be formed by fusion. Fig. 5.12 indicates that the areas which make up Group I cluster in the top, left-hand corner of that graph – in other words, they are likely to have a higher proportion of persons in the lower Social-Classes and have fewer households centred on married couples and thereby more persons either not yet married or in the contracting stages of the life-cycle (i.e. pensioners). All these areas are central and are mainly composed of pre-1919 houses. The great exception to this are the inter-war council estates, virtually all of which appear in this group. It is likely that this may be because these estates still had very old populations in the mid-1960s.[33] Geometrically, Group I forms a cross with arms stretching out into Belgrave, Humberstone, Aylestone and Evington.

If one looks at the sub-groupings within Group I (i.e. the thirteen level), these too can be associated with fairly distinctive characteristics. A and B split the areas in Group I about a dimension of housing quality. A is made up of the older areas, where the properties have a poorer standard of amenities, there are a higher proportion of unfurnished tenancies and a greater number of householders in the lowest social groupings. They are, for the most part, closer in to the city centre. B is only marginally different in character. The rest of the pre-1919 districts in the city are mostly in this grouping. They are mainly the slightly better-quality dwellings built between 1885 and the First World War, and include such areas as North Belgrave and Aylestone Park which were prosperous, lower-middle suburbs in 1914. C is a quartet of areas in the city which have the oldest age-structures, and includes both private and council areas, with two of the oldest local authority estates, dating from the 1920s, being included (Areas 54 & 70). The last sub-grouping, D, very clearly represents those areas undergoing redevelopment in this period (Areas 6 & 37) and have both very old populations and low Social-Class scores.

Group II

Group II forms a linear dimension on Component 1, indicating that the life-cycle/household-structure element does not play a great role in distinguishing it. This is a group dominated by local authority housing, and therefore it is not surprising that they predominantly score negatively on Component 2, which measures the Social-Class dimension. (The one exception is Area 63.) So far as the sub-groupings are concerned, F and G split about age-structure. F has the older populations and its four areas have a substantial number of specially constructed old peoples' dwellings. By contrast, G defines those council estates built in the 1950s which contained a large number of families in the expanding stage of the life-cycle. E is a most interesting sub-grouping because it is composed of three, rather different areas which have in common a mixture of old people and purpose-built flats.

Group III

Group III is easily identifiable as that which contains the areas with the larger and younger families in the city. The newest council estates are in sub-grouping I, whilst H is composed of an area of the inner city, which has been especially associated with the settlement of Asian families, and where there has been a literal rejuvenation as houses, before occupied by older people, have been bought up by younger families.

Group IV

Group IV is identified through its relationship to Component 2, only one area scores negatively on that component, and then only marginally. These are the better-off/middle-class areas of Leicester and are associated with owner-occupation and houses built after 1919. It does contain some areas on the borderline (e.g. Areas 16 & 28) but these date from just before the war, and were among those areas where sales by landlords seem to have been concentrated in the inter-war period. Sub-group J contains the bulk of the areas, and is composed of the great mass of owner-occupied dwellings built between the wars. L consists of the same sort of areas built after 1945. Because so much of the private development since the war has taken place outside the city's boundaries, there are only four areas in this grouping. K is another special grouping which picks out a set of areas in transition from one type to another. These are the areas on the periphery of the city's multi-occupied sector of housing, in which this sort of property has been expanding rapidly in recent years. Some of the areas (like Areas 28 & 59) are of very mixed housing character, others (like Areas 60 & 62) could eventually move completely into the multi-occupied sector.

Group V

Group V represents the core of the multi-occupied/furnished lettering sector and it is interesting that only three areas (Areas 20, 21 & 22) actually emerge in this group. These were the areas where multi-occupation was already spreading before the war and they are today the districts where this form of housing dominates.

Conclusions on the components analysis and classification

The function of the components analysis was to try to examine more comprehensively the relationship between housing and the main social and demographic dimensions in the city. That household structure and social class emerge as the main factors is hardly surprising given the variables put into the analysis, but the integration of these two components, as attempted by the cluster analysis, is especially interesting in terms of the character of the housing market.

If one considers the upper levels of the linkage tree (i.e. the five groups), one must conclude that the most fundamental element in the classification structure is social class. Groups I, II and III are primarily negative scorers and IV is distinguished by its monopoly of positive scorers on Component 2. Group V is a more complicated case. It is a very distinctive group, as indicated by the fact that it survives as an independent element in the linkage tree till so late in the fusion process, although there are only three areas in this group. It has links with both Group IV, which contains areas 'half-way' towards this group, and Group I, which is much more like the score pattern for this group.

This initial observation typifies the manner in which the housing structure intervenes between the two 'fundamental' elements of household composition and social class and thereby modifies and to some extent confuses them. Although, one might recognise social class as being the most basic dimension in the construction of the linkage tree, it is also clear that for the thirteen sub-groupings, the family-cycle/household-composition element plays the major role in differentiation. The connection between the two strata is provided by the special position of housing. Of the thirteen sub-groupings, only C does not possess a fairly definite housing character (it may therefore not be accidental that C is the last of the sub-groupings, apart from the multi-occupied zone, to be fused with any of the others).

In summary, there seems good reason for assuming that housing plays an intermediate role in the city's spatial structure. The contemporary housing market has converted the simple, direct relationship between housing and social structure that existed in the nineteenth century into a triangular interaction between social structure, demographic grouping and housing tenure. This new relationship is not completely consistent as indicated by the blurred edges of the linkage tree groupings, but it would be extraordinary if it were so. Leads and lags must be introduced into the city by the dynamic character of the urban social and spatial systems. Hopefully, more knowledge can be gained of this by the further examination of the pattern of mobility and migration in the Leicester of the 1960s.

Mobility in the Leicester of the mid-1960s

There was only one available source of information on movement in the Leicester of 1870 – the Borough Electoral Rolls. Today, one may reinforce such information by the material provided by the Enumeration District statistics of the Census. Before the pattern of movement provided by the Electoral Rolls is examined, these ED figures will be used to lay down the basis of mobility in the city.[34]

The Five-Year Mobility Tables of the 1966 Census were based on a question asking persons on the 24 April 1966 if they had had a different address five years before. 28.4 per cent of the population of Leicester aged over five years old answered 'Yes'. This was a smaller proportion than in either the whole of the East Midlands Region – 30.5 per cent, or England and Wales – 33.1

per cent. The total number of persons who had moved was composed of three elements. Over half the movers (17.6 per cent of the city's population) had moved from somewhere else within Leicester, under a quarter had come from somewhere else in Great Britain (7.5 per cent of the population) and about a tenth (3.3 per cent) from abroad. The last is a reflection of the importance of Commonwealth immigration as a factor in the city in the early 1960s, since it is nearly twice the average figure of 1.8 per cent for England and Wales.

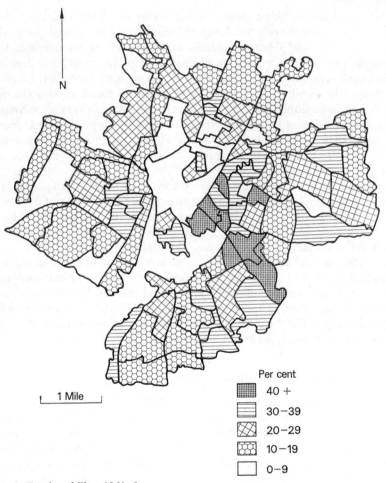

N

1 Mile

Per cent

40 +

30−39

20−29

10−19

0−9

Fig. 5.13. Total mobility, 1961–6

Fig. 5.13 shows the total percentage of persons in each of the eighty-one areas who had had a different address five years before. The most striking characteristic of the city has become the variation in rates of local mobility. In the most mobile districts, the rate of movement was five times that in the least mobile, and this must be compared to the position in 1870. Indeed, the comparison is more striking when one realises that the average annual rate of movement in the 1870s was two-thirds that of the quinquennial rate in the 1960s. The spatial variation in mobility supports some fairly obvious hypotheses. The highest rates of all in the city are associated with the multi-occupied districts (e.g. especially in the inner, south-eastern sector). Every area which had over 40 per cent of its occupants having moved in the previous five years was in this south-eastern sector. Overall, the inner areas have higher rates of mobility than the outer. The components analysis suggested that such inner areas were associated with smaller households, and especially with older people. If the relationship of mobility and age is to hold good it follows that the description of such areas as being associated with smaller households

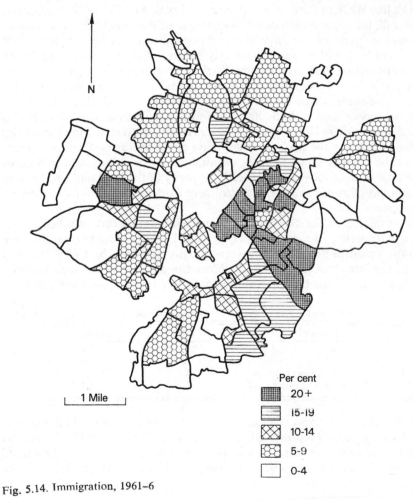

Fig. 5.14. Immigration, 1961–6

must include persons in both the earliest and latest stages of the family-cycle model. This tends to be confirmed by the fact that family-type households are the ones most associated with the newer, post-1919, outer areas. The outer areas generally are characterised by differences between the local authority estates and the owner-occupied areas. The former almost uniformly have the higher rates of mobility.

Fig. 5.14 breaks down total mobility by looking at simply that group of people who moved into the city from elsewhere. Again the variation between areas is considerable, and indeed, variations in the total level of mobility appear to be strongly related to differences in the rate of immigration between areas. The multi-occupied sector of the market is especially important here, and is picked out not only in the south-eastern sector of the city but also in a district like Western Park, where there is an isolated outlier of such property.

Overall, the combination of intra- and inter-urban mobility differs between the various parts of the city. The multi-occupied areas have high rates of movement both within and into the city. So to a slightly lesser extent do the post-1919 private areas in the outer districts, though in these parts of the city, the relative importance of inter-urban movement is greater than for intra-urban. The same is true for the areas of the inner city which are known to have been associated with the immigration of Commonwealth Asians in this period, e.g. Spinney Hills and South Belgrave. By contrast, the local authority housing estates on the city's periphery have about average rates of movement within the city but very low rates of immigration, which tend to pull down their total rates of movement. Indeed, for many of the local authority estates the amount of movement coming into them from outside the city is virtually non-existent.

The explanation for the pattern of mobility almost certainly lies in two areas, both well evidenced in other studies. The first is that there are basic relationships with the life-cycle/career-mobility model such as were sketched out in the introduction. Table 5.2 shows the association between mobility and age-structure for Leicester as a whole. Only broad age-groups are avail-

TABLE 5.2   *Mobility and age-structure, 1966 Census (percentage moving in previous year)*

Age-group	Males	Females
1–14	10.6	11.4
15–44	15.9	14.6
45–retirement	4.6	4.7
Post-retirement	4.2	4.3

able, but the lower levels of mobility found for older people in both national and local surveys is reproduced. Table 5.3 does the same thing for similar broad socio-economic groupings in the city, and it too confirms the overall tendency for movement to be more likely the higher up the social scale one goes.

150

TABLE 5.3 *Mobility and socio-economic group, 1966 Census (percentage moving in previous year)*

Professional workers	23.7
Employers and managers	12.7
Skilled manual workers	9.8
Clerical workers	13.x
Semi-skilled workers	11.0
Unskilled workers	9.9

Note: This table refers to the mobility of adult males who were economically active at the time of the 1966 Census.

It is difficult to see just how these city-wide associations fit into the urban housing pattern and the social geography of Leicester, but fig. 5.15 goes some way to providing some general indications. It takes the eighty-one areas and plots them by the five main groups that the cluster analysis placed them in against their mobility (as illustrated in figs. 5.13 and 5.14), both for movement within and into the city. It suggests that the interrelationships between the life-cycle/career-mobility model, housing structure and the pattern of the city's social geography are extremely complex. The relationship between mobility and social class is perhaps the most easily determined from the graph and the original maps. More expensive housing areas do have higher rates of mobility than the lower, and this is illustrated by comparing the results of movement for the post-1919 owner-occupied areas with that for the local authority estates. But life-cycle and housing structure elements are much more difficult to highlight. The inner areas seem to be characterised by higher overall mobility, yet it has been shown that they contain a disproportionate number of older, smaller households which do not move frequently. Of course, the specialist multi-occupied areas are characterised by a very young and very mobile population but they only form a small minority of inner city areas. For the rest, there are enormous local variations. The redevelopment areas have very low mobility (which is logical if one assumes that whatever populations they had at the time of the Census were waiting to be rehoused by the local authority), yet opposed to these are other very mobile districts of the inner parts of Leicester (the most extreme cases of which are in the immigrant settlement areas of Spinney Hills).

The spatial pattern of mobility does suggest a broad relationship between tenure and mobility. Furnished letting has very high rates of movement, owner-occupation moderate rates and council housing low, with unfurnished letting in a somewhat indeterminate position. But the last should sound a note of warning about drawing such conclusions from any areal pattern. The inner city districts are no longer solely associated with unfurnished letting. The division between pre- and post-1919 areas on the basis of tenure was probably relevant in 1939, but no longer holds good today, with the loss of unfurnished tenancies to owner-occupancy.

Both national and local surveys have suggested a rather different relationship of tenure and mobility with owner-occupation and council housing

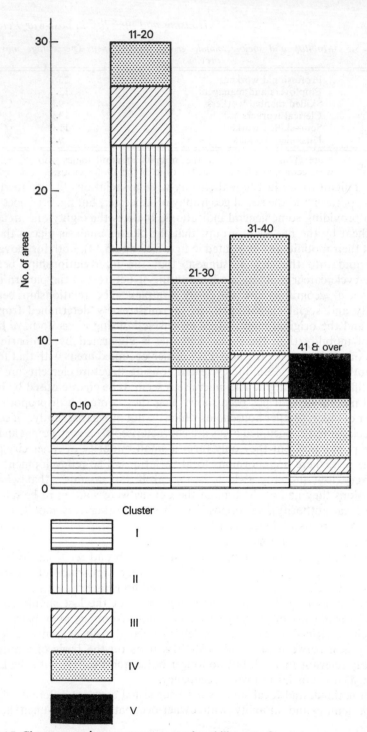

Fig. 5.15. Cluster groupings: percentage total mobility over five years

having rather similar levels and being contrasted with a higher rate of movement in *both* tenanted sectors of the market.[35] Such a pattern is by no means
inconceivable in Leicester if one accepts the mixed character of the inner city
areas, since it is possible that owner-occupied dwellings in those areas contain
the immobile, older households, whilst the let sectors, unfurnished as well as
furnished, contain the mobile. This contains some disturbing implications.
It implies very strongly that one cannot in such inner areas of the city recognise
direct and simple relationships between socio-demographic characteristics
and the housing structure, at least not so far as they have had a discernable
effect on the city's spatial structure. That would be true at a point in time,
but this is a study concerned with the dynamic aspects of the social geography
of the city, and it is a sensible hypothesis that the pattern which may be
observed in the city today may represent a snapshot of an on-going process of
social and spatial change. To begin to consider this, the pattern of movement
in the city will now be discussed.

The intra-urban migration pattern

The same basic methodology as in the study of 1870–1 will be used to look
at the contemporary pattern of movement; that is, a cross-sectional analysis
of intra-urban migration will be examined through investigating changes in
Electoral Rolls. Modern Electoral Rolls are far more comprehensive than
those found in the nineteenth century. Progressive extensions of the franchise
meant that by the middle 1960s, all British and Commonwealth citizens
(including citizens of the Irish Republic) resident in Great Britain and over
21 years old were entitled to vote.[36] Registration occurs annually in October,
usually by means of a general delivery of forms to householders. Registration
is not compulsory and some people do not bother, or in other ways get left
off the Rolls. Recent evidence has suggested that there must always be some
degree of error in their compilation (Dunn & Swindell, 1972).[37]

Information was collected for the year 1963–4 (actually from 10 October
1963 to 10 October 1964) and, for the purposes of the study, the Leicester
area was divided into the three parts shown in appendix B. Zone A consisted
of the City of Leicester as of this date, and for this area as complete an analysis as was possible given the normal constraints of time and information was
attempted. As with the 1870–1 Rolls, the basic process of collection was
through the compilation of losses and gains between the two successive years
and a subsequent procedure of cross-checking. The cross-tabulation of name
lists has both advantages and disadvantages when compared to the same task
in the nineteenth century. The most obvious added difficulty is one of increased scale. The Rolls for the early 1870s had less than 15,000 names, those
for 1963 over 180,000. But against that there is the much smaller chance
of error in the analysis produced by the greater detail of names given
and especially the results of adult suffrage which mean that all members of a

household over 21 are recorded and thereby the probabilities of error in comparing moves are drastically lowered in most cases.[38] Again in Zone A, as in the city of 1870, the records of the Registrar-General were used to identify the role of deaths and marriages in the migration system.

It has already been shown that a considerable amount of the post-1945 growth of Leicester occurred outside the city boundaries, and Zone B took account of this. It consisted of that penumbra of built-up land immediately adjacent to the city and physically indistinguishable from it on the ground. Within this peripheral zone, a complete check of Electoral Rolls was made, but because of the difficulties caused by these areas being in different registration districts, it was decided to treat this area solely as a 'sink' for local emigrants from Leicester.[39] Zone C is equivalent to the area covered by the Leicester Telephone Directory, and is thus approximately the same as the rest of the county area. For this third area, the Telephone Directory was used to provide a preliminary search procedure designed to cover the emigration of households from Leicester to those commuter villages which have been especially important in the spread of the city's influence in the last few years. This preliminary investigation was then followed up by checking the destinations hypothesised from the Directory against the appropriate entries in the Electoral Rolls for that area.[40]

The Electoral Roll compiled on the 10 October 1963 contained 181,264 individual entries. The total number of rateable hereditaments in the city in 1964 was 84,990, and when this is related to the findings of the 1961 and 1966 Censuses, there must have been about 90,000 Census-type households in the City in that year.[41] Over the year 1963–4, the city's Electoral Roll had 19,521 losses and 18,614 gains within Zone A. This is a percentage loss of 10.8 and a gain of 10.0 per annum. The overall net loss was 907 which reflects the declining population of the central city in the late 1950s and 1960s.[42]

There are again three reasons for losses – death, movement within the city and emigration. Of the 19,521 individual losses, 2325 were identified as deaths. This is only about 70 per cent of the total recorded deaths in the city in this period (1963: 3558; 1964: 3338). The shortfall is almost certainly due to the deaths of long-term (i.e. over six months) residents in institutions like hospitals who were not included in the study and the few deaths of persons under 21 (though these only account for about 8 per cent of deaths in 1963–4, compared with about 40 per cent in 1870–1). All told, 669 deaths seem to have resulted in a vacant household (which is only about a quarter of the estimated number of changes proportionately produced in 1870–1). Of the remaining 17,196 losses, 15,029 were in multi-person households who moved.[43] There were 7157 such households or about $8\frac{1}{2}$ per cent of the city total. 3860 of these could be traced to addresses elsewhere in Leicester (i.e. in Zone A), and 3297 left the city, of whom 816 went to addresses in Zones B and C. 346 moves by multi-person households apparently resulted in their amalgamation with another household in some dwelling. There were there-

154

fore left 2167 individual movers. Many were single-person households, but there is also the special problem of marriage. The situation here is more complicated than in the nineteenth century. Male halves of marriages can simply be traced through the Rolls directly, what is needed is some measurement of the female halves where persons simultaneously change both their names and their addresses. As a result of the search of the marriage registers, 1088 women marrying were identified, 742 of them to men who also moved, 346 to men who stayed at the same address. In 1963 and 1964, there were 2388 and 2389 marriages in the Leicester Registration District, respectively and therefore the identified total is only 45 per cent. There are three main reasons for the difference – persons marrying and then leaving the city, persons marrying and not changing their address (e.g. continuing to live with parents) and undoubtedly of the greatest importance, women marrying under the age of 21.[44] There is therefore left one final component – single person household moves – of which there were 337.[45]

Gains to the 1964 Rolls are more easily considered. Out of the total of

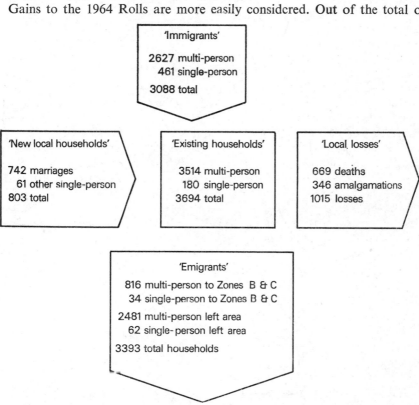

Fig. 5.16. Diagrammatic simplification of the city's migration system, 1963–4. (The diagram plots only household moves which cause or fill a vacancy in the stock of housing in the city, and thereby ignores some of the individual changes in the electoral rolls.)

155

18,164, 2301 were apparently new voters reaching 21. These included all 'Y' voters and any other persons joining households with the same surnames as themselves.[46] There were 7585 new households, 3694 of them the result of moves within Leicester, 803 apparently newly established but locally recruited, and 3088 immigrant households.

The total pattern of movement is diagrammatically simplified in fig. 5.16, whilst table 5.4 compares the major parameters of the system as derived from the Electoral Rolls with the situation shown by the One-Year Migration Tables of the 1961 and 1966 Censuses. It is comforting to see that there is an overall agreement between the Rolls and the Census, at least so far as these

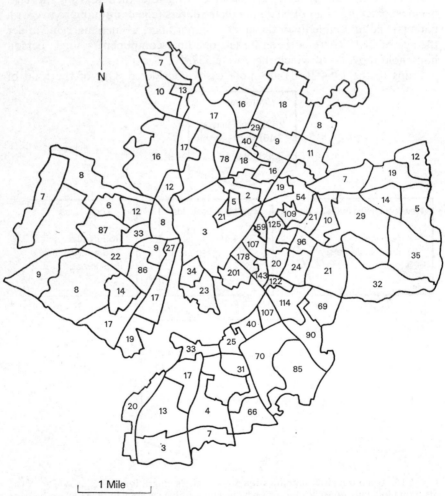

Fig. 5.17. Immigration, 1963–4: numbers of incoming households (total 3088)

TABLE 5.4   *A comparison of the rates of mobility derived from the 1961 and 1966 10%
sample migration data of the Census and the Electoral Roll*

	Within city	Emigration	Immigration	Total
	%	%	%	%
1961 Census				
Persons	6.2	4.7	3.0	10.9
Households	4.4	3.5	2.1	7.9
1966 Census				
Persons	6.2	3.7	2.7	9.9
Households	4.5	3.1	2.0	7.2
1963/4 Electoral Roll sample				
Persons	5.6	3.9	3.1	9.5
Households	4.2	3.6	2.9	7.8

Note: All figures are annual rates of movement.
Total mobility is measured as the total proportion leaving the address in which they
were resident a year previously, and is therefore the sum of the within-city and emigration
columns.

broad household mobility rates are concerned, and this gives added confidence to the analysis that follows.

Obviously, the central problem in analysing the information collected
from the Electoral Rolls is its sheer size. As a first step, the evidence for
immigration and emigration is presented in Figs. 5.17 and 5.18, respectively.
The latter includes those households which went to locations in Zones B and
C, and therefore is composed of both short- and long-distance movers.
Both maps reinforce the patterns of the Census, with immigrants especially
concentrated in the multi-occupied districts and certain special areas like the
district of recent Asian settlement in Spinney Hills. The better-quality,
private, post-1919 areas also have numbers of such migrants, whilst the
local authority estates have *very* low proportions of immigrants and low
proportions of emigrants.

Even excluding immigration and emigration, though, there are over 5000
individual moves. Figs. 5.19 and 5.20 show the first steps in analysing this
material by presenting the pattern of movement at the ward level. Fig. 5.19
shows gross flows, fig. 5.20 the net. Gross movements tend to emphasise local
movements between neighbouring wards, but certain points do suggest
themselves. In particular, there appears to be a nexus of movement in the
older, inner parts of the city which bears some resemblance to the pattern of
closely linked movements typical of the mid-nineteenth century. Outside this
inner ring of wards, a much stronger directional bias becomes apparent.
Generally, outward flows are stronger than inward. To draw a familiar
analogy, the city's migration pattern is like a wheel: the hub is formed by the
closely interlinked inner city wards, its rim by the succession of local authority
estates and post-1919 private developments, and the spokes by the outward
flow of people from old to new, cheaper to more expensive, private letting to

157

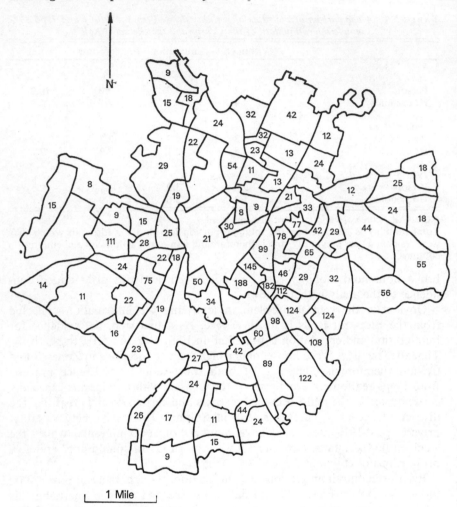

Fig. 5.18. Emigration, 1963–4; numbers of outgoing households (total 3393)

council or owner-occupied property. This is well demonstrated by the net pattern. The central hub tends to disappear because the interacting moves between the inner wards cancel one another out. The spoke element, on the contrary, is greatly strengthened. Inner wards such as Latimer, Wycliffe and the Castle all experience substantial net losses, whilst wards on the periphery like Evington, de Montfort, Aylestone and Newton receive immigrants from these inner wards. It is especially interesting that the intermediate wards with locations straddling the centre–periphery division (and generally a more mixed bag of housing), like Westcotes and Spinney Hills, also occupy an intermediate position in the city's migration structure, taking in movers from

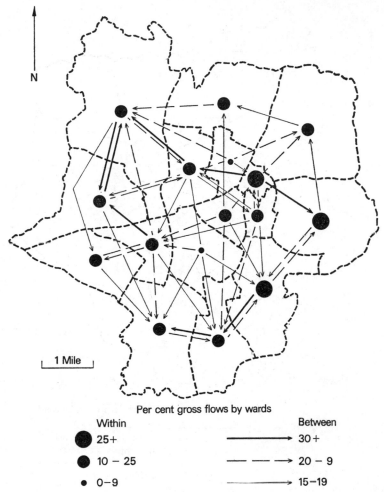

Fig. 5.19. Gross movements at the ward level

the central wards whilst still contributing a net outflow to the peripheral areas.

At the level of the sixteen wards, which are poorly structured on all but the most general criteria of age and housing, little of substance can be gained. Therefore, an attempt to make more sense out of the structure of movement will be made through the use of the principal components technique already used to describe the socio-geographic structure of the city in the middle 1960s. A technique of this type is necessitated by the size of the matrix produced by the eighty-one divisions used in that previous analysis. Within the city, there are approximately 6500 cells (i.e. an 81 × 81 matrix), but it is

159

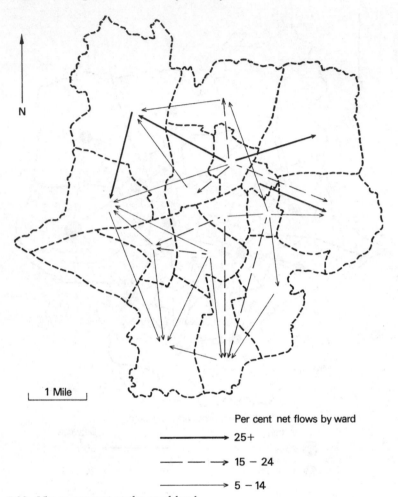

N

1 Mile

Per cent net flows by ward

→ 25+

─ ─ → 15 − 24

───→ 5 − 14

Fig. 5.20. Net movement at the ward level

also wished to include those areas immediately adjacent to the city (Zone B) which have been divided into ten local authority areas, and a generalised area approximating to Zone C, which contains the rest of the county. There is therefore an asymmetrical matrix of the size 81×92.

The use of a principal components analysis to identify general patterns of linkage in a spatial system has become quite common (e.g. Illeris & Pedersen, 1968; Goddard, 1970; Clark, 1973), but the components produced by this form of the analysis differ in interpretation. They are less likely to be related to any general dimensions in the structure, because of the non-linearity produced by the inevitable number of zero cells, but rather to groups of areas biased towards one end or other of the spectrum of all possible values (i.e.

160

instead of getting a range of loadings or scores running from positive to negative, the tendency is to produce an association of strongly positive values opposed to a great mass of poorly differentiated cases centred about the neutral position of zero). Each component produces a pattern of loading which represent the origin areas for that component and a set of respective scores which represent its destinations. In each of the following maps which illustrate the components, arbitrary levels of significance have been set at + or − 0.4 for the loadings and + or − 1.0 for the scores.

Component 1

The most significant component produced by the analysis explained 13.69 per cent of the total variance. It identifies (Fig. 5.21) an element which previous chapters have made familiar. This component covers a horseshoe-shaped area encompassing the southern part of the inner city. The areas identified as origins take two types – late-nineteenth-century, working-class terraced districts and those areas which have become associated with multi-occupancy in recent years (Areas 20, 21 & 22). There must be a close degree of interlinkage in this component, since most of the areas also appear as destinations. This is very much the southern part of the 'hub' seen in the ward analysis. There are also some areas which appear exclusively as destinations. They are predominantly post-1919 in age and owner-occupied in tenure. Mixed areas do appear (e.g. Areas 78 & 79 in North Braunstone) but no district exclusively devoted to local authority property.

It is interesting that this component stretches right across the inner areas on both eastern and western sides. Not only must there be considerable amounts of local movement in such wards as Westcotes and Spinney Hills, but these areas must also be linked together. There are spatial constraints, however; Belgrave, which is a very similar area does not appear on this component, and similarly the better quality areas of Knighton are excluded, as is Aylestone to the south-west. This is not a component associated with middle-class areas. Rather the manner in which the closely interlinked inner wards are feeding people out into post-1919 housing seems evidence of something very akin to the filtering process which operated before 1914. In the inter-war city, the barrier between the pre- and post-1919 cities, emphasised by the differences in tenure, rent controls, style of property and so forth has been emphasised. This component straddles that barrier and perhaps indicates that it has been breaking down in the post-war period. Certainly, this would tie in with certain of the processes which seem to be implied by the general analysis of the social geography. It has been seen that a large number of properties in these areas have gone from tenanted to owner-occupation, so in that respect the barrier no longer exists. It is also true that the inner areas by the early 1960s were beginning to be characterised by a demographically mixed population of young and old, and the movement across the 1919

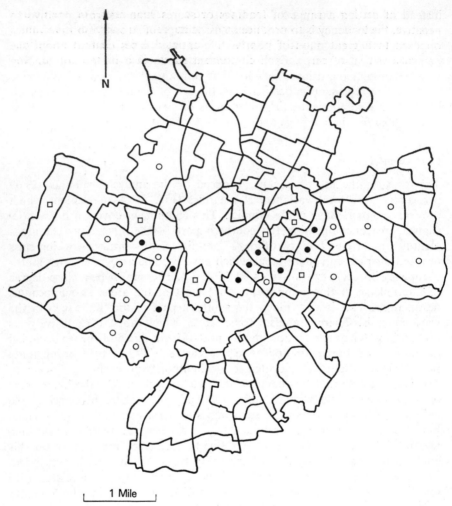

Fig. 5.21. Migration component 1: origins and destinations
Key: □ origin. ○ destination. ● both

division may indicate an example of 'career-mobility' whereby as families expand they seek more modern accommodation.

Component 2

By comparison to the wide extent of Component 1, Component 2 (fig. 5.22) which accounts for 10.46 per cent of the variance, is a much more restricted feature of the city. It is very clearly centred on the council estates of West Leicester. All three estates – Braunstone (especially that part built after 1930),

162

New Parks and Beaumont Leys are represented. The high degree of interaction within and between such areas is indicative of the general policy of the council in encouraging tenants to seek as good a fit as possible between their dwelling type and size and their household circumstances. One of the commonest

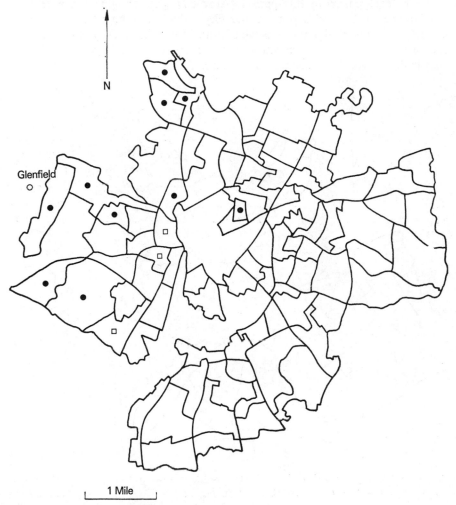

Fig. 5.22. Migration component 2: origins and destinations
Key: □ origin. ○ destination. ● both

examples of this is the way in which older people, when their household size shrinks, may be encouraged to move from the standard three-bedroom family dwellings to specially designed old peoples' residences which can be found on almost every estate.

Opposed to this important amount of movement within the estates is their significant association with a smaller group of pre-1914 areas. This undoubtedly represents the source of one of the major flows of new tenants into council property, especially since one sees here an area which was undergoing redevelopment in this period (Area 37). One should also note the appearance of Glenfields outside the city, which was an area where the council was undertaking extensive building in 1963–4. Only one area to the east of the River Soar appears in this component, and not surprisingly that is the St Matthews council estate (Area 3).

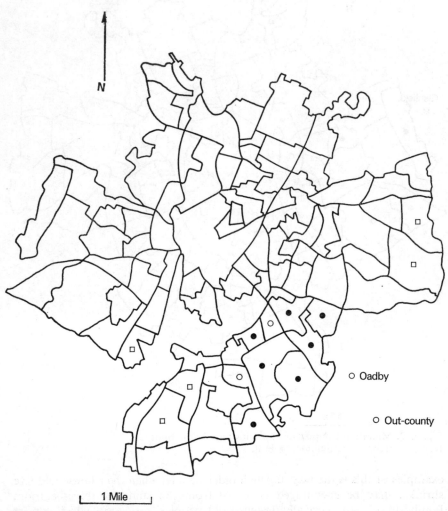

Fig. 5.23. Migration component 3: origins and destinations
Key: □ origin. ○ destination. ● both

Component 3

Component 3 (fig. 5.23 and 8.56 per cent of the total variance) once more identifies that most familiar of all the elements in the city's structure, the high-value housing sector in the south-east of the city. This area is centred about Knighton but includes some of the cheaper, flanking areas of owner-occupied housing in Aylestone and Evington which appear as origins on this

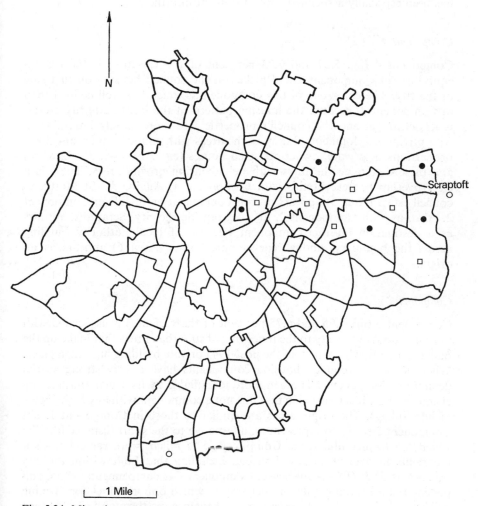

Fig. 5.24. Migration component 4: origins and destinations
Key: □ origin. ○ destination. ● both

165

component. The most important destinations for this component are Oadby and the generalised outer-county Zone C. This still seems likely to be the component which identifies middle- or even upper-middle-class movement. The traditional high-value sectors within the city, especially the older parts like Stoneygate, have been losing the better-off families to such areas outside the city ever since the end of the war. Though this component indicates some attraction of persons from the moderate-value flanking areas, the most significant input to the areas within the city in this component has probably been the conversion to multi-occupation, but this does not show up because it has been especially associated with movement into the city.

Component 4

Component 4 (fig. 5.24 and 7.73 per cent of the variance) is the eastern equivalent of Component 2 and picks out the council housing on that side of the city. The core-zone of this component is much less well defined than in Component 2, because the housing pattern of the local authority estates is less clear cut and the council's properties are more closely intermingled with private developments. Many of the areas with council estates also have proportions of private, owner-occupied houses (e.g. Areas 48, 51, 52 & 56) and these determine the links with neighbouring private areas. There is a clear overlap because of this with Component 1. Again, the St Matthews council estate (Area 3) appears on this component as it did on Component 2. Also, as with Component 2, an area outside the city (Scraptoft) appears as a destination on this component because of local authority building across the boundary, and a redevelopment area (Area 6) appears as an origin.

Component 5

Component 5 (fig. 5.25 and 6.75 per cent of the variance) is associated with movement out of the city to the inner ring of suburban areas that make up the bulk of Zone B, where most of the post-1945 private building has taken place. Whereas the two council housing components have a very strong spatial definition, Component 5 is much more identifiable by its social and housing characteristics. It is unusual in that it is not a component with any high degree of interlinkage. The origin areas are similar to those in Component 1, but Component 5 seems to represent an alternative to the short-distance filtering which the former illustrates. Component 5 is much more radical since it represents movement to new housing, including that built within the city (e.g. Areas 57 & 69). In one sense, Component 5 is complementary to Component 1 since it represents another way in which opportunities for opening up the mobility system of the city present themselves through links between the inner and outer parts of the housing structure.

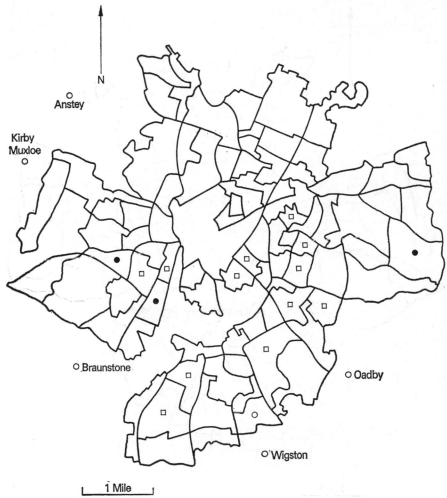

Fig. 5.25. Migration component 5: origins and destinations
Key: □ origin. ○ destination. ● both

Component 6

Component 6 (fig. 5.26 and 6.51 per cent of the variance) picks out a distinctive and 'isolated' part of the city's migration system associated with the northern part of Leicester centred on Belgrave. This has always been a morphologically very distinctive section of the city forming a narrow sector between the River and the old Midland Railway. Crossing points over these barriers are limited, and the sector represents a reasonable cross-section of the city's housing structure from the very earliest terraced developments of

167

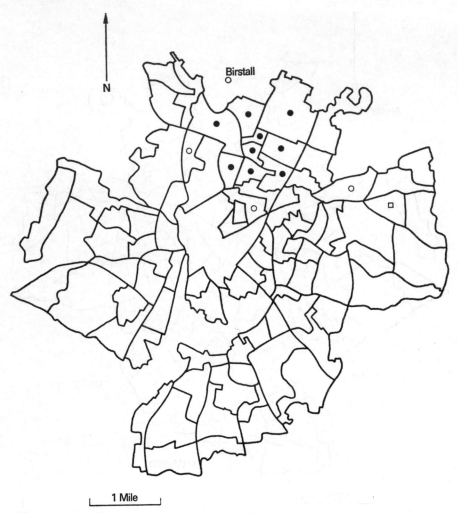

Fig. 5.26. Migration component 6: origins and destinations
Key: □ origin. ○ destination. ● both

the 1870s just north of the Great Northern Railway line through the inter-war houses of North Belgrave to the modern developments beyond the boundary in Birstall. Belgrave is a microcosm of the city's development and the migration pattern is a reflection of this. It is spatially and functionally still similar to that which seems to have existed before 1919. It is worth emphasising here that this supports the contention that changes in the structure both of the migration system and housing market in the last fifty years represent a modification and addition to what went before rather than a complete revolution.

168

Component 7

Component 7 (fig. 5.27 and 5.71 per cent of the variance) is the last of the extracted components and is also associated with a spatially distinctive area, Aylestone. This is also morphologically distinctive, but it is especially interesting that this component is the only one in which there appears to be a strong link between private and council-owned housing. This is specially true of the moves made to areas outside the city, into both Glen Parva,

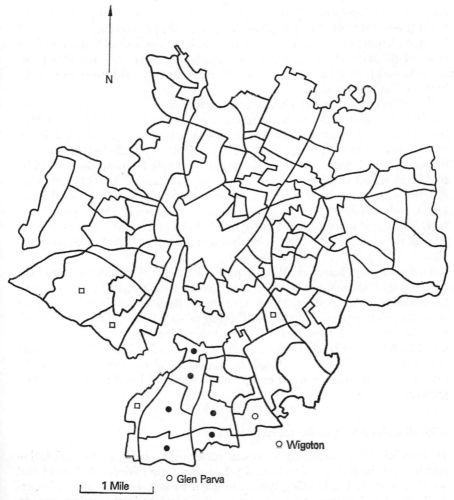

Fig. 5.27. Migration component 7: origins and destinations
Key: □ origin. ○ destination. ● both

169

where there was recent council building, and Wigston, where there was extensive private construction in the post-war period. Also, again, recent areas of private building in the city appear as a significant destination (Area 69).

It is extremely difficult to know where to halt the component analysis of flows. The situation is somewhat different from the structural analysis. Here, lower order components pick out quite distinctive, if successively smaller and more tightly drawn districts of interlocking movement. They are easy enough to interpret, but the loss in generality with succeeding components causes a process of diminishing returns. The seven components so far extracted account for 60 per cent of the total variance, which is a very satisfactory result given eighty-one 'variables' and the somewhat discontinuous nature of the data. Component 8, not illustrated here, identifies a closely linked set of movements in North Braunstone and marks a progression towards components which pick out relatively small areas already subsumed within the higher order components.

The principles upon which the components seem to be constructed are twofold. The first relate to basic features of spatial correspondence, which have always been common factors in the development of the city. Many moves remain short-distance and locally orientated, and there is nothing unusual about Leicester in this. The West Midlands Housing Survey (1971, pp. 51–3) found that 36 per cent of all moves were made within walking distance (compared to 76 per cent within the boundaries of one local authority), and nationally 28 per cent moved such a distance (Cullingworth, 1965, table 28). Local movement is most typical of the inner areas of the city, and least of the outer, middle-class areas, but it is partially valid of all parts of the city.

This has always been true. What is particular to the migration system of the modern city is the manner in which the current housing situation has modified the traditional local pattern. There have always been associations between the structure of migration and the distribution of social class. They were dominant in the city of 1870 and form the foundation for the analysis of this chapter. What is different about the post-1945 situation is the disaggregation and spatial confusion introduced by the lack of correspondence of the modern housing pattern when compared to the traditional pattern of urban growth.

Migration, housing and the social geography

In examining the obviously complex relationship between the migration pattern and the housing situation, fig. 5.28 uses a simplified structure of that housing market. Six housing types are used. Firstly, council properties are easily identified because they are recorded separately in the rating records. Secondly, mutli-occupied dwellings were defined on the simple principle of

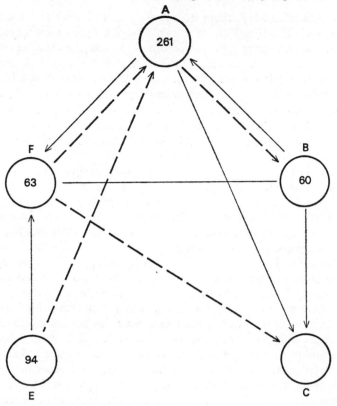

Fig. 5.28. Movement within the housing market, 1963–4
Key: A. private under £70. B. private over £70. C. private outside. D. commercial.
E. council. F. multi-occupation. Numbers within circles indicate movement in tenures

taking all properties where there were more than two adult males with differing surnames.[47] Thirdly, all those rateable hereditaments jointly described as residential and businesses were picked out, and such properties as shops with integral flats have been described as commercial. The rest of the private dwellings were split into three groups. The fourth category were all those new private dwellings outside the city boundary in Zones B and C to which Leicester residents had moved.[48] Within the city, the private properties were divided about the £70 per annum rateable value. This level was selected for two reasons. Firstly, it allows comparison with the city's total rateable structure as described in the Annual Financial Reports of the City Treasurer and secondly, £70 does reasonably approximate the significant division between the pre- and post-1914 housing.[49]

Unfortunately, it is not possible to distinguish between tenanted and owner-occupied dwellings in the private sector from the data here available, but one may make certain assumptions. In 1964, Leicester had 88,900 domestic hereditaments. 2400 of these were 'commercial', i.e. combined residential and business premises and 25,900 were owned by the local authority. Of the remaining 60,600 private dwellings, 43,000 were rated at less than £70 per annum. The 1961 Census identified 23,000 dwellings rented unfurnished from private landlords, but by 1966 (with a slightly larger area), this had fallen to 19,700. Given what has already been said about the pattern of building since 1920 and the concomitant distribution of rateable values, it is a fairly acceptable assumption that the great majority of these were rated at less than £70. But this must mean that by 1963–4, the majority of houses under £70 per annum were owner-occupied and is indicative of the rate at which such older, cheaper properties have been converted from tenancy to owner-occupation in the post-war period, since it has been stressed that the bulk of these pre-1914 properties would have been built originally to let.

Fig. 5.28 illustrates the total flows between and within each of the six categories. Taking each in turn:

A. *Private dwellings rated at less than £70 per annum*

With 43,000 dwellings, this is the largest of the six categories. In all, there were 2440 households who moved into or within this category (5.7 per cent of the total). Overall, there was a net loss of 404 households from this grouping, two-thirds of them through the redevelopment schemes in progress in this year (i.e. Areas 6 and 37). Within the city, this category is a net loser of households to all the other groupings except multi-occupancy. No less than 829 households moved within the grouping, a fifth of all the moves within the city. The strongest connections are with council housing; 615 households moved from this category of housing to local authority properties (though only 166 moved back). It is very interesting to consider the relationship of this clearly important element in the flow pattern of the city to the spatial analysis already carried out. Nearly half the moves into the council sector are produced by the redevelopment schemes, and these links do appear strongly in that previous analysis, with each of the redevelopment areas being linked to the group of council estates on either side of the city. Yet it is interesting that the spatial analysis did not produce any high-level component associated with movement into the council sector as a whole. The explanation probably lies in the alternative means to redevelopment by which people obtain council properties (i.e. by accumulating sufficient priority in the local authority's waiting list), which means that each year a scattering of households will move into the local authority sector from older, cheaper houses all over the inner areas of the city. Relatively, this component of housing attracts far more immigrants to Leicester than it contributes emigrants from it. There are twice as many of the former as there are of the latter. 46 per cent of all the immigrants to the city come to dwellings in this category.[50]

B. *Private dwellings rated at more than £70 per annum*

There are about 13,500 dwellings in Leicester in this category. It is more mobile than the cheaper properties of group A with 12.1 per cent of its households moving in a year. This is to be expected in terms of the career-mobility element and is borne out by the spatial pattern of mobility which showed that the owner-occupied areas of post-1919 housing had people who moved more often. Movement within the group is not more frequent than the average with 340 households relocating within the same sort of properties in Leicester. Similarly, with 1290 households leaving and 1284 entering, there is a nice balance which reflects the fairly stable state of this housing group, with new building in the city (e.g. in Area 69) being counter-balanced by some conversions to multi-occupancy or to commercial use in areas nearer the centre. Within the city, these more expensive private houses are net gainers from all other sectors, and especially from the cheaper private group. One should also note the strong linkage with private housing outside the city boundaries in Zone B and C. Overall, immigration and emigration to and from the city are very important for this category. No less than 36 per cent of all emigrants from the Leicester area came from this category of housing, which is $2\frac{1}{2}$ times what one might have expected from the group's share of the total housing stock.

C. *Private housing in Zones B and C*

For the purposes of this analysis, this type of housing was treated only as a recipient of movers. 757 households were traced from Leicester as moving into this form of housing (another 93 went to local authority properties outside the city boundary). Of the immigration into this category, 341 (45 per cent) came from the cheaper private housing in Leicester, 273 (36 per cent)

173

from the more expensive private housing, 76 (10 per cent) came from the multi-occupied sector, and only 67 (9 per cent) from the council sector.

D. *Commercial properties*

There is little that can be said about this small element in the city's housing market, other than generally immigration and emigration balance and that there is a limited flow between it and all the other sectors of the market.

E. *Local authority housing*

There are nearly 26,000 local authority dwellings in Leicester and household mobility is very low at only 4.5 per cent per annum. Considerably more households moved into this category than left it, but over 40 per cent of all moves are within the council sector itself. Relatively, local authority housing attracts net migrants from the cheaper private housing and multi-occupancy, but generates net movement out to the more expensive private dwellings in the city and the private property in Zones B and C, but it must be acknowledged that the movement to these latter sectors is minimal both in the particular migration system of the city and the council sector in detail. The particular feature of the local authority sector of the market is the very low proportion of immigrant households. Only 6 per cent of all immigrant households to the city are accounted for by a tenure that now takes up nearly 30 per cent of the total dwellings stock of the city. Immigration to places outside the Leicester area is relatively more important at 12 per cent of all emigrant households, though this is still considerably lower than the scale of this sector would lead one to expect.

F. *Multi-occupancy*

The most extraordinary feature of multi-occupation is its mobility. Though it is much, much smaller than the council or more expensive private sectors, it produces about as many movers within the city as either of these two. Household mobility in this tenure appears to run at about 33 per cent per annum.[51] About 40 per cent of these moves are actually within the multi-occupied sector which argues for a very fluid situation within such areas. Multi-occupation is also important in terms of movement into and out of the city.

Fig. 5.29 illustrates the particular effects of one specialist aspect of migration behaviour in the housing market – those flows within the city which occurred either at the time of marriage or otherwise when setting up a new household. These are the flows implicit in those 803 households which were locally recruited in Leicester during 1963–4. One can see that the cheaper

174

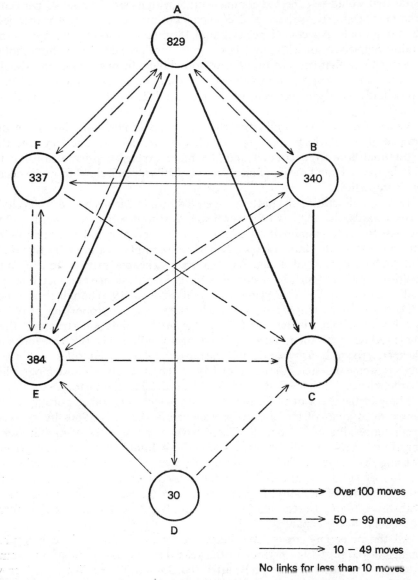

Fig. 5.29. Movement as the result of the formation of new households
Key: A. private under £70. B. private over £70. C. private outside. D. commercial.
E. council. F. multi-occupation. Numbers within circles indicate movement in tenures

private housing of the market recruits about that proportion of new households that would be expected commensurate with its size – about 43 per cent. The more expensive private and council housing sectors recruit rather less than one would expect – 11 per cent and 18 per cent respectively. Again, the multi-occupied sector with only about 5 per cent of the city's total households receives twice that proportion of new households. In particular, one should note the special importance of the new housing in Zones B and C. 117 new households (nearly 15 per cent of the city total) located themselves in these areas.

Generally, the flows implicit in the setting-up of new households are the reverse of those which apply for established households. One can note the substantial flow from council and the more expensive private housing to the cheaper private housing, and there is a similar 'black-flow' to multi-occupied properties.

Earlier in describing the pattern suggested by movements between the city's sixteen wards, the analogy of a wheel was used to describe its structure. One can see from the combination of the social geography and the analysis of movements that the 'hub' area is a much more complex feature than the 'rim'. The 'hub' is composed of an intermixture of cheaper private housing and multi-occupied properties. It contains virtually all the privately let housing and almost every dwelling in the city built before 1919. Though these inner areas have a demographic structure which has a great proportion of older, smaller households, in terms of their role in the migration structure, they are associated with the movement of younger families. The two categories of cheaper, private housing and multi-occupancy play similar roles within the city's movement system, though they have rather different socio-demographic characteristics and have had a different recent residential history.

Many of the areas of unfurnished letting, which was still important in the lowest-rated areas of the city, were occupied in the early 1960s by an ever-declining residue of old people. These have very low rates of mobility and contribute to the overall lower rates of mobility found in the cheapest private housing. By contrast, the multi-occupied sector is occupied by a high proportion of young, unattached households and therefore is associated with very high rates of mobility. Irrespective of their household tenure, movers in both these kinds of areas tend to be of similar character in that they form households in the expansion stages.

All the inner city areas have been undergoing a cycle of demographic change in the post-1945 period. In the years between the wars, and to some extent in the years immediately after the Second World War, they grew steadily older in population as well as housing as the very low rates of mobility that developed after 1914 stabilised their populations and gave most areas a demographic uniformity that had not existed in anything like the same form before the First World War. Many of the householders in such areas were the same in 1945 as they had been in 1919. Such a process cannot be demographically infinite because of the natural span of human lives. In the 1950s, there began a process of rejuvenation. In its most extreme forms, this was seen in the rapid changeover of population found in the rapidly growing multi-occupied sector and in the special alterations in demographic structure brought about by the immigration of Asian households into such areas as Spinney Hills and South Belgrave. Generally, the process grew into a polarisation of such areas' populations into young and old and then into their age structures growing younger as the older element 'wasted away'.

The process has continued instead of merely setting off on the cycle again because the new populations inhabiting these inner city areas have been far more transitional. They have seen their occupation of dwellings in such districts as a stepping stone to something better – either a council house or a newer, owner-occupied property (perhaps outside the city's immediate boundaries). Of course, this is a generalisation disguising a lot of complications at which this analysis only hints. There is a clear distinction between the unfurnished and furnished tenancies which is reflected in the latter being much more attractive to those who may never wish to occupy a permanent dwelling in Leicester, as for example students. Similarly, in referring to those inner areas as a transitional stage, one must acknowledge that some people pass through it much more quickly than others.

The choice presented to persons in the inner areas who wish to seek a better standard of accommodation (and most of the more modern and comfortable dwellings are outside them) is one primarily determined by income (or perhaps more accurately by potential income). The difference between the contributor and recipient areas is mainly based on the family-cycle element. It is not a young-old division. Rather young and old are associated with the contributor areas and the middle-aged (using that term in a very wide sense) with the recipient. In comparison, the contrast between the two recipient categories, council housing and owner-occupation, is the contrast of the career-mobility element. One cannot state with complete confidence that all who can afford it choose owner-occupation in preference to council property, though it seems reasonable to assume that most do. That very few council tenants move across into the more modern private property is not perhaps surprising given the much higher standards of modern council housing and the difficulties of raising a mortgage.

The return movement in the system is provided partly by the location of new households, but as much by the differential location of immigrants and emigrants to and from the city. For the better off, this is comparatively unimportant and the sector of more expensive private housing is able to operate outside the main body of the city's housing market with a high degree of movement in and out of the city being as important as the movement within the city. For the less well-off, however, the closing of the local authority sector to most immigrants is of fundamental importance, and is the main mechanism behind the centrifugal system of household movement and the converse centripetal system of new household formation.[52]

Perhaps the most interesting feature of the present relationship between housing structure, social geography and movement system is the obvious fact that this relationship cannot be a static one. The 'system' that exists today is very much a function of an interaction between events in the housing market (such as the introduction of council housing), fundamental alterations in behaviour (such as the fall in mobility during and after the First World War) and the working through of the processes that such events implied in the context of the city's existing housing pattern. The changes that have occurred since 1945 with the progressive rejuvenation of many inner-city areas are indicative of the sorts of processes which are inherent to the present social geography. For example, natural processes will almost certainly ensure the progressive ageing of most of the council estates. Some of the inter-war estates, e.g. Saffron Lane, are already quite 'old' areas. Conversely, extensive local authority building in central area redevelopment schemes would lead to these areas becoming relatively 'young'.

Similarly, one may observe the probable effects of the erosion of private letting and its almost total replacement in the private sector by owner-occupation. In the inter-war period, the division, even 'barrier', has been emphasised between the let and owner-occupied sectors. Since 1945, that 'barrier' has been breaking down and the importance of Component 1 in the analysis of movement stresses the strength of the interaction across the '1919 barrier'. If present trends were to continue and owner-occupation to become the sole private tenure (with the possible exception of a specialised furnished-letting/multi-occupied sector), then it would be possible to envisage a return to a system of movement within the private sector, very similar to that existing before the First World War, with the system being composed of a series of overlapping areas melting into one another as they climb the economic and social scale. There would be differences, and such differences lead one into the final conclusions as to the major changes that have occurred in the migration/housing relationship and the influence which they have exerted on the social geography.

CHAPTER 6
CONCLUSIONS

There are two stages in which the conclusions to the study may be considered. Firstly, the major changes which have occurred in residential mobility in Leicester over the past century will be outlined and their causes discussed. Secondly, the more general case of housing as a factor in the spatial evolution of the city can be examined, especially so far as the example of Leicester is typical of urban areas in Britain.

Mobility and housing

The central fact that has been stressed in this study is that over the past hundred years, there has been a marked fall in the number and frequency of household moves in Leicester. Though there have been significant short-term variations, on average mobility in the 1960s was only about 40 per cent that found in the 1870s. Whereas a century ago, 1 in every $4\frac{1}{2}$ households could be expected to change their address in any year, today that has been reduced to about 1 in 11. The overall fall in the rate of residential mobility has been modified by two conditions. Firstly, the reduction in mobility has been much more conspicuous in the poorer districts of the city and among those families in the lower social groupings. There has been a reversal in the relationship of mobility and social class such that a negative association in the nineteenth century has been converted to a positive one today. Secondly, though there is evidence of a secular fall in the rates of mobility, the reduction was most marked about the period of the First World War. The war itself produced unprecedented strains in the city's housing market which were reflected in the collapse of the high rates of movement characteristic of the pre-1914 period. Although there was a recovery after 1919, mobility rates have never achieved such levels since. It was also during the war that the change in the relationship of mobility and social class was most marked. These simple conclusions suggest three sets of hypotheses which may account for them. These might be termed the urban growth, life-cycle/career-mobility, and housing market explanations. Strictly speaking, they are not so much alternatives as complementary approaches to the overall problem which highlight differing aspects.

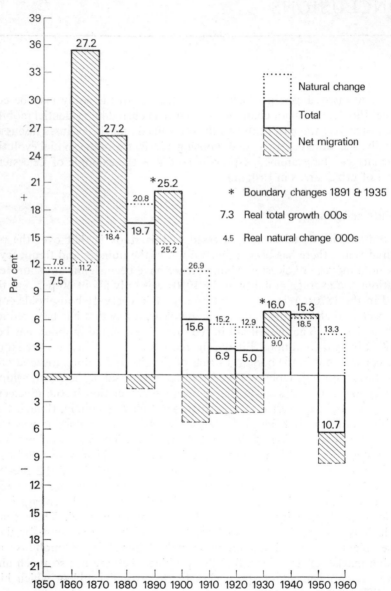

Fig. 6.1. Natural change and migration in Leicester, Oadby and Wigston, 1851–1971

Urban growth

Robson (1973, pp. 96–7) has discussed the manner in which migration reflects population and economic growth, and it is an attractive idea that the fall in mobility reflects those circumstances in which proportional growth of the urban population has been much slower. In the chapter on movement in the nineteenth-century city, it was emphasised that rapid growth must have formed the base for a continuing resorting of the urban population, and lower growth might be expected to exert a damper on mobility if only because of the smaller pecentage of new households coming on to the market annually.

Fig. 6.1 illustrates the demographic growth of Leicester, Oadby and Wigston from 1851 to 1971, splitting the growth into the results of natural change and net migration to these areas.[1] Real and proportional growth is shown. The falling-off in growth is clearly illustrated. For natural change, the peak decade for growth was 1901–11 in real terms and 1881–91 in proportionate terms. For net migration to the city, both in real and proportionate terms, the decade 1861–71 was most important. On average, growth per annum has fallen from a maximum of 3.5 per cent in the 1860s to 0.5 per cent in the 1960s.

There are two main objections to this view. The first is that one is actually concerned not so much with the growth of population as with increases in the number of households in which growth has been much more marked. Taking just Leicester, between 1871 and 1971 the city's total population

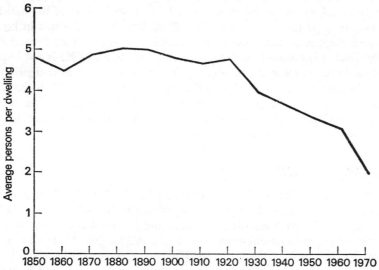

Fig. 6.2. Changes in the average size of households in Leicester, 1951–71. (*Source:* Census.)

increased by approximately 3.0 times, whilst the number of households in the city increased by 4.8 times.[2] Fig. 6.2 shows changes in the average size of households between 1851 and 1971 and the marked fall since 1930 is very obvious. Partly, that fall has been associated with lower fertility and smaller families (Marsh, 1958, pp. 41–5), but as important has been the rise in head-ship rates common to the whole country.[3] In recent years, the rate of house-hold formation has far and away exceeded the rate of population growth (Donnison, 1967, pp. 30–3). Special evidence of the increase in headship rates is the growth of one- and two-person households, which increased from 26.8 per cent of all households in 1931 to 53.5 per cent in 1971. Such rises in headship rates are accounted for by a range of changes that have encouraged the fragmentation of the family which are themselves indicative of the second objection to the urban growth hypothesis. Though as a general argument for encouraging a greater amount of stability in the urban population it is quite acceptable, the mechanisms by which lower mobility would have been achieved do not tie in completely with observed evidence. The pattern of movement into and within the city has altered substantially but not always in the manner which this viewpoint might suggest.

Immigration to the city has altered very much in the direction and scale that would be expected. Taking the figures from the two cross-sectional analyses suggests a minimum that might apply in the two periods; of immigra-tion running at about 40 per cent of total mobility in the city, which suggests that there is some consistent association between immigration and the total level of residential mobility. Opposed to this is the observation that within the city, the decline in movement has especially been associated with a falling off in those highly localised moves which were so typical of the mid-nine-teenth-century pattern but which only effected marginal changes in housing costs and standards. Such movement remains of significance (especially in the multi-occupied sector) but the structure recognised in 1963 is one imposed by moves between tenures and housing sectors which do appear significant in terms of changes in costs, quality and condition. Parallel to this is the way in which the differences in mobility have become much more marked between parts of the city. All this suggests that any explanation needs to take some account of the internal dynamics of the city.

Life-cycle/career-mobility

The character of this model was outlined in the introduction, and some attempt has been made through the directories to investigate changes in its structure, especially for middle-class areas and occupations. It was suggested there that changes in the life-cycle have had considerable affects on the social geography. The most obvious change has been the manner in which later or 'contracting' stages of the life cycle have become longer. Hole and Pountney's investigation showed how whereas a woman marrying in the 1860s was likely to have her last child reaching the age of 15 when she herself was 57, for a woman marrying in the 1950s, the comparable age would be 44 (Hole & Pountney, 1972, pp. 15–18). When one adds this to the longer life expectancy of persons today, it means that in the mid-nineteenth century the post-children stage might have been twelve years on average, today the comparable period could be thirty years. The whole numerical and proportionate import-ance of older households in the city has increased, such that households headed by persons of pensionable age have risen from 15 per cent in 1871 to nearly 40 per cent in 1971. It is difficult to specify how much of the fall in overall mobility has been due to such changes. The section on the movement of the occupational samples suggested that over and above such develop-ments, there had been a clear and observable decline in rates of movement at all ages. Certainly, the patterns of social geography seen in the middle 1960s display a clear significance for age-structure as a major differentiating feature in the city.

The role of the career-mobility element has been also of some importance. It has been described how it is likely that there have been changes in the relationship of social class to migration to and from the city, and whereas in the nineteenth century such long-distance movement was proportionately more important for the poorer groups, today, that relationship has been reversed. The city is no longer growing through the rural–urban migration of cheap labour but rather by differential movement between urban areas. It is a process in which prosperous areas like Leicester attract a higher propor-tion of the career-orientated, middle-class occupations in which economic growth in recent years has been concentrated. By contrast, the need and the desire to move on the part of persons in less well-paid occupations has declined. This is especially reflected in the low rates of inter-urban mobility found in local authority housing, though here the rules of local authorities in allocating houses may also be of significance. This last point illustrates the way in which both life-cycle and career-mobility changes have become of significance through their translation and magnification by parallel changes in the urban housing market.

Housing market

The direct influence that changes in housing structure and housing policy have had on mobility seems to have been fairly obvious. The relationship between the housing market and the likelihood of movement within and into the city has been seen in terms of shorter fluctuations such as those that were especially prominent before 1914. But longer-term changes have also been of significance. In general, the last fifty years has seen the progressive disaggregation of the unified Victorian housing market into a series of sub-divisions based on tenure. It has been suggested that private letting as a tenure was especially conducive to movement within the city because it allowed frequent, fine adjustments of supply and demand in housing costs. It also minimised a householder's attachment to his property both financially (and probably) emotionally. The costs of moving were sufficiently lower to encourage tenants to seek to maintain the balance of their housing costs with their incomes, and the more indirect balance of their total locational costs by frequent movement. This frequent adjustment nevertheless meant that most moves only effected small-scale changes in individual circumstances.

For many, such a flexible system was mandatory in the nineteenth century because of their inability to cope with fluctuations in income levels and the marginal nature of expenditure on housing. So little spare income was available in the nineteenth century and the protection against the calamities of life was so minimal that movement to maintain an equilibrium between revenue and costs was absolutely essential. It is for this reason that mobility was negatively associated with income and social class and the frequency of movement related in the minds of social reformers with poverty.

This century has moved away from this situation. Not only have incomes risen to allow better standards of accommodation, but more importantly most people have much better protection against violent fluctuations in income. The most obvious case has been the increased ability of pensioners to maintain separate households. The reasons for this are manifold. Old age pensions are the most obvious factor, but they are only one part of the set of rebates, subsidies and protection for the householder that the welfare state provides. In direct housing terms, the provision of subsidised council housing has been the most important feature, and it is no coincidence that the fall in mobility in the lower social groupings should have been paralleled by the growth of this tenure.

Changes in the administration of the private tenures have been as significant. Rent controls and the role of legislation which decontrolled property with vacant possession have been a major factor in dramatically reducing the level of mobility in many older areas of the city. They have been accompanied and have been closely related to the switch from tenanted to owner-occupied property. This changeover has been yet another factor in reducing mobility. It is true that present movement rates in owner-occupied areas are often higher than in other parts of the city, but they are still nowhere near the rates found in all areas of the nineteenth century city. Owner-occupation will discourage movement in a number of ways. Firstly, it makes moving more expensive relative to the potential advantages to be gained from relocation. Legal fees, estate agents' charges and so forth mean that such direct costs of moving have risen to the point where they may equal some months' costs of occupation. Owner-occupiers also have a direct interest in maintaining their property (for many of them it is their biggest single asset, after all) and in the last few years, do-it-yourself and home improvement have become commonplace. The most extreme cases of such action, which now occur very frequently, are the numerous home extensions that seem to be springing up on every private estate. Instead of moving to meet new demands, the owner-occupier is just as likely to try to refit his existing property.

Of course, there is still a place for the sort of property which caters for a transient population, and the spread of the multi-occupied/furnished-letting sector has met that need. It is in this sector of the contemporary housing market that the conditions most like those of the city of 1870 exist. It is the only part of the present city where there are levels of mobility comparable with those typical of the nineteenth-century circumstances. It is also typical of these areas that the pattern of movement is very reminiscent of the nineteenth century with its predominance of short-distance moves. In these areas, one has a population who often have many of the problems which were so common before 1914 – low incomes, an inability to raise credit, a susceptibility to unemployment and general loss of earnings.

At a more general level, one can recognise in recent years how the evolution of the housing market in Britain has paralleled and encouraged the development of such external economies as important factors in the spatial structure of the city. This has especially been true in the case of owner-occupiers, where the coincidence of interests between owners and occupiers which was so rare in the nineteenth century has made opposition to any form of detrimental change so strong. In the nineteenth century, the landlord was usually faced with a *fait accompli*. If the area in which his property was located declined for some reason, he was more than likely to find that his tenants had upped and gone, and the tenants themselves had only a limited interest in preserving the values of their landlords' properties. Today, the situation is quite different. The owner-occupier has a direct and often vital interest in protecting the value of his own property, and the rash of amenity societies, residents' associations, local action groups and so forth illustrate this.

All three hypotheses as to the decline of mobility in the city have some significance, and the really very major changes in the workings of urban areas in Britain that have been demonstrated here, have very much been the product of a combination of all three. But it is the third set – those changes reflecting specifically on the organisation and workings of housing – that are by far the most interesting in that they have given British cities a distinctive geographic character.

Housing and the spatial evolution of British cities

All the evidence of this study has reinforced the initial premise that housing represents a uniquely interesting topic in the geography of the city because it plays an intermediate role in its spatial organisation. 'Above' housing there are those factors such as 'technology' which are common perhaps to a civilization. The most obvious example in any city's spatial structure has been the place of transportation systems. Accounts of urban development couched in terms of a progression from 'pedestrian' to 'mass-transit' to 'automotive' cities have been very common in geography, and one should certainly not underwrite the significance of such innovations. Most generally, they represent one example of the way in which the complete paraphenalia of industrial society determines a basic structure which will underlie most twentieth-century cities.

One can equally recognise features which are 'below' housing in scale, the most obvious being all the individual characteristics of particular sites. Leicester is not a city which has had its growth predetermined by physical features, but as in all cities, they have played a role (the growth of the high-value, residential sector in the mid-nineteenth century is perhaps the most obvious example in Leicester's case). Also of importance have been the conditions of economic and social change in the city. The character of the city's industrial base, its continuing prosperity, the role of immigration over a very long period, all have given Leicester a particular character.

In looking at housing in Leicester, one might recognise five somewhat arbitary and artificial stages, which form 'ideal-types' which may be used to illustrate the role of housing in the urban spatial structure and the manner in which residential mobility has operated within that housing structure.

Pre-industrial (*? before 1820*)

The circumstances of the pre-industrial city have not been examined in detail in this study, but they obviously form its preface. Though the pre-industrial city may not have been distinguished by a lack of residential mobility, the overall social geography of the city was much more stable. Social segregation did exist, but it was combined with, and sometimes subordinated to, functional segregation (Vance, 1967). Whatever mobility took place was as likely to act in a counter-balancing fashion with little net result on the urban spatial structure.

Early-industrial (*1820–65*)

From the end of the Napoleonic Wars, an ever-rising tide of immigrants completely altered the socio-demographic base of the city. Unbridled urban expansion occurred with the minimum of public control and builders had to meet the demands of a poorly paid workforce by erecting hundreds of new dwellings to the lowest possible standards. Urban conditions deteriorated, and site values became of much greater importance. Even the rich and powerful were unable to maintain the conditions of these areas in the centre of the city where they had traditionally lived and they deserted the area for new dwellings on the periphery. The need to maximise the utility of land and the restrictions imposed by a city still organised for the pedestrian meant that growth only accreted very slowly at the edges of the built-up area.

Conditions of drainage and water supply emphasised the positive correlation between altitude and site values.

The social geography of early Victorian Leicester reproduces many of the conditions found by Engels in the Manchester of the 1840s. The whole city is reminiscent of Marx's analysis. Social segregation developed very rapidly, but it was still unsophisticated. Middle- and working-class districts became increasingly differentiated. The city was extraordinarily fluid but formless and chaotic. Households were continually on the move, especially in the poorer areas, but most moves were short-distance and seemed to accomplish only the most marginal of changes in housing standards or conditions. Immigration was very important, but an enormous number of people also left the city in any year, and when trade was bad (as in the 1850s) emigration could actually exceed immigration. The particular way in which Marx's analysis seems especially appropriate was in the direct relationship between the economic organisation of the city and its housing market. The rationale of the migration system of the city was the need which most families seem to have felt to maintain an equilibrium between their effective level of demand and the quality of property they could afford, which allowed little leeway in bad times.

The 'ecological' city (1865–1914)

Between 1860 and 1880, there was a fundamental change in the way in which Leicester grew. Most obviously, perhaps, it was associated with the shift from a dependence on foot transport to a mass-transit system (Warner, 1962), but equally if not more important was the evolving relationship between the city's 'maturing' industrial economy and its social structure, which was reflected in parallel changes in the pattern and standards of new dwellings brought about by an increasing public control of development which stemmed from the growing public concern with urban health and hygiene. Leicester was a city which was still growing rapidly in the last quarter of the nineteenth century and it displays exceptionally well all those features that might be expected as typical of the classical 'ecological' models. As Timms has stated, 'the Burgess scheme of urban growth and structure was developed to fit the pattern of rapidly growing cities, with industrial bases, efficient transportation, heterogeneous populations, free-market housing conditions and a value system that stressed newness and spaciousness' (Timms, 1971, p. 222). Of all these conditions, only that of a heterogeneous population was missing from the Leicester of this period. With rapid population growth, the city expanded very quickly and it is in these last years of the nineteenth century that the familiar patterns of sectoral growth and a generally concentric distribution of property values are easily recognised. Rising incomes, especially in the upper-working and lower-middle classes were accompanied by ever-rising standards of new building, standards which were increasingly backed up by public action.

Implicit in such patterns and their growth concomitant with the city's growth was an equally rapid process of residential change. With continuous adjustments in locational costs and benefits and thereby changing residential values, successive waves of property followed one another, and the outward movement of social groups continued. Class and social status seem to have become much more complex phenomena, and much less easily defined than in the city of the early industrial period. Housing came into its own as a symbol of rank and position and the rental structure of the city's stock of dwellings became infinitely more involved. But social change did not occur in any process of fits and starts, rather it was a continuous process with the speed accelerating and slowing in time with the cycles of building activity and periods of tightness and slack in the city's housing market. Frequent mobility was the natural associate of this system with its periodic gluts and shortages of accommodation. Over the whole period from 1865 to 1914, each wave (and there were such waves in the early 1870s, the middle 1880s and after the turn of the century) pulled out further social strata from the older, cheaper, poorer-quality properties of the central area, and stretched out the great mass of the working- and lower-middle-class population over a social and rental gradient running from the slums of the centre to the new suburbs of the 1890s. For most of the population in this period, movement as the alternative to the manipulation of spatial economies remained dominant, but already one can see how in the middle-class areas of the south-eastern sector, some evidence of the development of attempts to influence these values did occur. Also, in a more general sense, the imposition of rising building standards through legislative and administrative action was a direct manipulation of the structure of costs and values in the market to the advantage of the middle-income groups and the detriment of the worst-off, who found that new property had been taken completely out of their reach. Success in the ecological city was by 1914 almost directly measurable by one's distance from the city centre (with the exception of course of the very best and most spatially distinctive families).

Inter-war (1920–40)

The First World War was an event of unparalleled significance in the housing history of British cities. Changes initiated by the war caused them to diverge radically from the predominating patterns of the previous forty years. Goheen (1969), in his study of Toronto, recognised the pattern of the contemporary city very clearly in the Toronto of the 1880s and 1890s. To some extent this must be true of all cities (if only because so much of the fabric of the city of that period still survives), but in Leicester and, it is submitted, most British cities the differences are as marked as the similarities. The war introduced a vastly increased amount of public intervention in the housing market through rent controls and local authority dwellings.

Mobility had collapsed during the war, and never really recovered afterwards. Some areas had only a quarter or a fifth of pre-war levels. The old system of social change that was recognisable before 1914 did not reappear and the city was characterised by the development of a real segregation not only associated with social class but also with stages in the life-cycle. This was true of both ends of the demographic spectrum. On the one hand, there was the emergence of newly built districts, both public and private, which on construction became inhabited by a population concentrated in the young family stages of the life-cycle, yet which because of the low mobility of households grew steadily older as the inter-war period progressed. The same process created at the other extreme, districts which had been new before 1914 into areas which were middle-aged by 1939.

The city in the inter-war period displayed a new pattern of spatial organisation in which there were no longer the simple direct relationships between income, social class and the costs and standards of housing. Many of the poorest families now occupied modern properties built to a comparatively high standard. The old system of mobility in which there had been a close link between movement and adjustments in the urban housing market broke down. The new properties which were built in the inter-war period were distinguished because they were of a different tenure from the one which had dominated the city's housing before the First World War. The new tenures, council housing and owner-occupation, had constraints to entry which weakened their links with the older let properties.

Post-war (since 1945)

Since the Second World War, many of the changes that had been initiated in the inter-war period have become magnified. Public control of development has been far more important and overall public involvement in the housing market much greater. It is in this period that the model of housing classes has emerged with full force. The urban housing market has been typified by a much greater degree of disaggregation with groups of properties being characterised by some particular set of legislative or administrative conditions. The main distinctions of such groups have been their association with tenurial classifications, in which the interactions between tenures have been only imperfectly articulated. Each tenure has been associated with a broad set of entry conditions that have become generally identified with groups of the population, but the composition of such groups has become much more complex than in either the pre-1914 or inter-war periods. Social class and income have continued to be major factors in residential location, but they have now been joined by stage in the life-cycle, potential career mobility and length of residence in the city. Also, such tenure groupings have not necessarily occupied a one-to-one relationship with any particular socio-demographic group. Some areas and types are much more mixed in character. For example, many of the late-nineteenth-century cheaper housing districts now have heterogeneous populations and neighbouring houses may contain families whose intentions and prospects are very different. Social segregation is certainly not disappearing, but it is perhaps becoming a more complex matter.

Such complications are reflected in the city's movement system. There are now very great variations both in the fluidity and role which individual districts play in the residential mobility process. Within tenurial groups, particular movement patterns predominate, as exemplified by the close linkages within the local authority housing estates, which are accompanied by specific relationships with those districts undergoing redevelopment, a general association with poorer housing areas in the city and very weak links with all areas outside the city.

In summary, one can recognise two interacting patterns of forces at work in the contemporary city. On the one hand, there are those, usually exogenously induced, changes in the structure and administration of the housing market which include policy decisions as to the location of new development, choices between such alternatives as redevelopment or improvement of older housing areas, and basic alterations in the legal framework of housing. On the other hand, there are the evolving demographic and social patterns such as are associated with the life-cycle/career-mobility framework.

Over a long period, one must ask the question as to how typical has been Leicester's experience. Certainly, in the late-nineteenth century, it was a well-housed city, but it is certainly a possible hypothesis that all English cities have been experiencing a process of convergence in this century simply because their local circumstances have been overriden to a large extent by the pattern of changes in the framework of housing superimposed by nationally determined and nationally applied policies. Even when the agencies of change have been local authorities, their actions have been predetermined to a large extent by central government.

Overall, the developing significance of housing suggests the considerable importance which has to be given to the political framework of the city in determining its social structure. The housing-class model, for example, is one which has only partial links to the economic basis of the city's social structure, and the way in which political action has interposed itself between the sort of direct associations that were so much clearer in the nineteenth century is very striking in contemporary British cities. Because housing is so much a politically influenced topic, it is undoubtedly the major means by which political considerations may be translated into the city's spatial structure.

The future

The implication of this study is that conditions must be expected to change in the future. The lesson of the last decade (since the detailed cross-section of the 1960s was taken for this study) is that the environment of housing continues to be in a state of flux. Not only are there long-term trends in the housing market, such as the continuing decline in the unfurnished letting sector, but also major policy developments. The 1969 Housing Act initiated a much greater emphasis on the improvement of older houses which has since been strengthened further. Security of tenure has been extended to the furnished letting sector and the conditions of this section of the market which so distinguished it from the unfurnished letting sector have been almost totally eliminated. The 1972 Housing Finance Act attempted a major revision in the financing of local authority properties, but has now gone and is to be replaced by something else. Sales of local authority dwellings to sitting tenants have become politically controversial, and the government has stepped in to subsidise mortgage holders beyond their present benefits through the taxation system. Every year seems to bring some new piece of legislation relevant to housing.

It is not the purpose of this study to speculate further as to what the results of these changes might be, or of what new developments the next few years might see. Cities are infinitely complex phenomena. All one can perhaps hope to do is to impose that form of explanatory structure which best appears to marry theory and practice from the standpoint of one's own position and era, and thereby to allow some sensible conclusions to be drawn concerning their main contemporary features.

DATA-SOURCES FOR INTRA-URBAN MIGRATION[1]

Unlike some European countries, Britain does not possess any form of continuous population registration.[2] Births, marriages and deaths are recorded, but virtually nothing is known about the movements of the population between these events.[3] It is true that there have been some officially sponsored statistics suitable as a basis for migration research. The National Register which was in existence during and after the Second World War did provide such a basis for a number of studies (Newton & Jeffrey, 1957), and the records of the old Ministry of Labour (Grebenik & Shannon, 1943) and more recently the National Health Service (H. R. Jones, 1970) have also been used.

Surprisingly, the Census showed relatively little interest in the migratory behaviour of the population until very recently. Though there have been a number of studies by historical geographers, using the county-based, birthplace statistics, which have reconstructed nineteenth-century inter-urban and rural–urban migration patterns (Darby 1943; Friedlander & Roshier, 1966), not until 1961 did the Census include tabulations based on a direct question as to previous residences. All the results from the 1961, 1966 and 1971 Censuses have been presented on a 10 per cent sample, and although they have provided a foundation for a new wave of work into inter-urban and inter-regional migration (Masser, 1970; Fielding, 1969), they have not allowed so much to be done at the intra-urban level. Coding has been restricted to the local authority and therefore use has been restricted to studies in larger conurbations (Johnston, 1969b; Moindrot, 1970; Chilton & Poet, 1973).

The consequent deficiency of data at lower scales remains a serious problem. Alternative sources are not readily available. In the United States, the records of public utility companies have been the basis for a number of studies (e.g. Boyce, 1969), but generally such resources are not available in Britain.[4] Most of the sources that have been developed in this country stem from limited, specially commissioned surveys. The best known of these is the General Household Survey, mounted by the Office of Population Censuses and Surveys to supplement the results of the Census itself. The scale of this survey, like all the others in this field, has been constrained by time and money. Indeed, Cullingworth & Orr (1969, pp. 143–6) have gone so far as to suggest that the resources necessary to produce a statistically significant number of recent movers are out of all proportion to the potential results.

Useful surveys do exist at all scales from the regional, as typified by the excellent set of papers on mobility and migration in North-East England produced by the University of Newcastle,[5] or 'Mobility and the North' (North Regional Planning Committee, 1967), down through the Department of the Environment's Conurbation Housing Surveys to the numerous local studies that have been produced. All such studies have concentrated on the reasons for movement, and have said relatively little about how movement fits nto any overall conception of the development of urban areas.

An exception that illustrates very well the efforts necessary to produce data relevant to such problems is Watson's study of residential movement in West-Central Scotland (Watson, 1971). Here, a special sampling frame was provided by OPCS and an extensive (and thereby expensive) back-up survey partially financed by the Scottish Development Department was required to produce the material for an investigation into the chains of movement initiated by new housing and their subsequent relationship to the process of 'filtering'.

LEICESTER'S ADMINISTRATIVE BOUNDARIES, 1835–1966

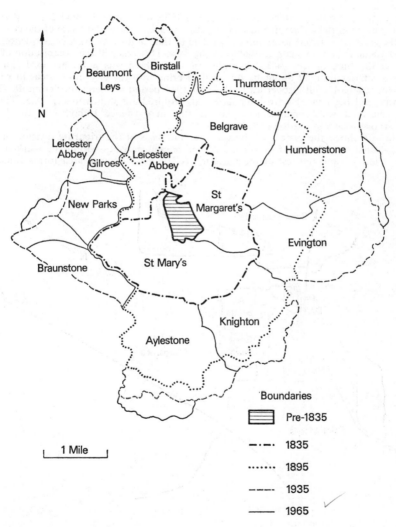

Fig. A.1. Administrative boundaries, Leicester, 1835–1966

Changes in boundaries, both for the city as a whole and for the sub-units of administration within it, are of some importance for the background to the study. Fig. A.1 shows the general expansion of the city since 1835. 1835 is a suitable starting point because it marks the Municipal Corporations Reform Act, which created the modern administrative arrangements for the city (Hartopp, 1933; Patterson, 1954). Fig. A.2 illustrates the within-city boundaries that existed between 1835 and 1892 (and that underlie the 1870 analysis). But anomalies soon developed as the great expansion of the city spilled over these boundaries in the 1880s. In 1890, the Leicester Extension Act brought into the city (as of 1 January 1892) another 5000 acres populated by some 34,000 people, previously living in the adjoining parishes of Belgrave, Knighton, Aylestone and Humberstone.[1] The extension necessitated a complete revision of the internal divisions of the city as shown in Fig. A.3.

A further boundary extension was made in 1935 with an additional 29,000 persons mainly living in the Evington and Humberstone areas on the eastern side of the city and on the great Braunstone local authority estate to the south-west. As a consequence, the ward boundaries were again revised as in Fig. A.4. This was still the organisation of the city at the time of the 1963–4 analysis. In 1966, there was a further extension to the boundaries brought about by the need to bring into the city those areas in which the council had built properties in the post-1945 period – Glenfield, Scraptoft, Glen Parva – and the revised ward boundaries this required are shown in Fig. A.5. This is the area and ward arrangement that is still in existence today, after the Local Government Reorganisation of 1 April 1974.

Fig. A.6 shows the system of areas used as the basis of the tripartite division of the 1963–4 analysis. The city forms Zone A; the named parishes within the marked line form Zone B; and the rest of the Leicester Telephone Directory area forms Zone C.

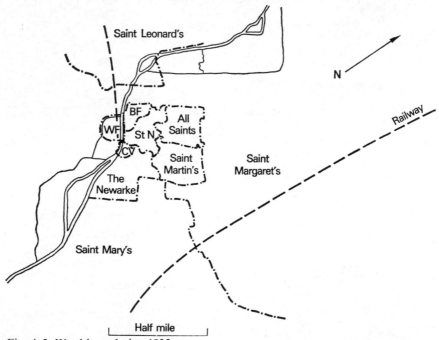

Fig. A.2. Ward boundaries, 1835
Key: CV. Castle View. BF. Blackfriars. St N. Saint Nicholas. WF. Whitefriars

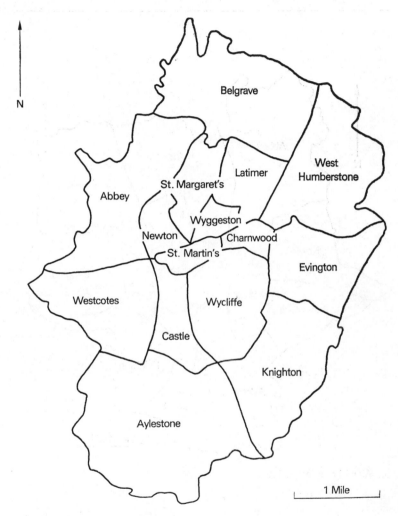

N

Fig. A.3 Ward boundaries, 1895

Fig. A.4. Ward boundaries, 1935

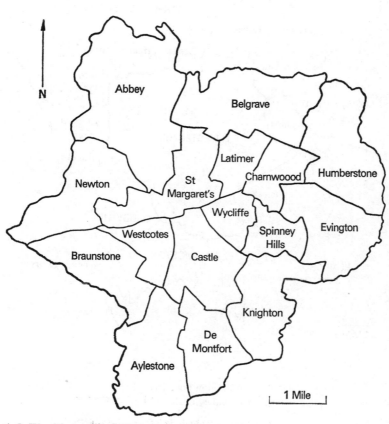

Fig. A.5. Ward boundaries, 1966

Fig. A.6. Areas of analysis, 1963–4

SOCIAL STRATIFICATION, 'CLASS' AND OCCUPATIONAL CLASSIFICATIONS: USES AND TERMINOLOGY

A central problem in this study is the relationship of geographic mobility to the social structure and the individual's position in that structure. Because 'social stratification' is such a key issue in sociology, it is necessarily a very complex problem.[1] Here, interest is focused on three issues – the terminology of the stratification system, the special meaning of the word 'class', and finally, the use to which the occupational classifications which the study employs may be put.

Parsons (1954, p. 69) has stated that, 'Social stratification is regarded as the differential ranking of the human individuals who compose a social system and their treatment as superior and inferior relative to one another in certain socially important respects.' It has become common among sociologists to recognise that the ordering of individuals within any social system is so complicated that the multiplicity of relationships cannot be subsumed within a single variable scale. In general, there has been a recognition following Weber of a tripartite division between the economic, prestige and legal aspects of social ordering – 'class', 'status' and 'power'.

Weber especially contrasted the economic basis of stratification that would produce 'classes' and the prestige ranking that might produce 'status-groups': 'With some over-simplification, one might say that "classes" are stratified according to their relations to the production and acquisition of goods; whereas "status groups" are stratified according to the principles of their consumption of goods as represented by special styles of life.' (Weber, 1948, p. 193.) This has tended to produce the division between American and European sociologists with the latter emphasising that social stratification and class structure are two distinct aspects of social organisation (Dahrendorf, 1959, p. 63). Weber's view has always been compared with that of Marx. Giddens (1973, p. 51) has stressed that although it is most common to characterise their differences in terms of the former's argument for a multi-dimensional view of stratification when compared to the latter's dominating idea that social relations are derived from the economic basis of society, a more general difference is the contrast of Weber's 'political paradigm' of social organisation to Marx's 'economic paradigm'. In the Weberian model of industrial society, a pluralistic model derives from a different and wider view of property and implies a special role for the administration of society, as recognised by Weber's emphasis on the place of bureaucracies in the evolution of modern capitalist societies.

The model of 'housing classes' is very much more related to the Weberian approach, especially so far as in modern Britain the intervening role of national and local bureaucracies has become so important in the housing market and in the allocation of individual properties. It also emphasises the severe syntactical difficulties that face anyone trying to make reference to the structure of any urban society and the place within it of such groups as 'classes'.[2] Both Marx and Weber have specific uses for that word, but it is much more difficult to account for the popular definition of such terms. 'Working class' and 'middle class' (plus all the intervening stages) are so deeply ingrained in the popular consciousness that one cannot hope to avoid them. This is especially so when one is faced with a situation where one is attempting to describe the structure of a society through something as limiting as an occupational classification.

There are two aspects to the problem of using an occupational classification, even if one accepts the basic fact that such a classification can only be a proxy for something much more complex. The first is that one has to have a clear conception of what one wishes the occupational classification to serve as a proxy for: much of the discussion of the use of occupational classifications has tended to become somewhat bogged down in a consideration of minutiae. It is obvious that no occupational classification can possibly hope to reproduce a strict picture of the sorts of class structures that either Marx's or Weber's use of that word entails. No such classification could provide sufficient data. What most occupational classifications seem to hope to do is construct a general model of the social structure which combines many elements.

This is very well illustrated by the sort of occupational classifications used by the Census, which are prepared by the Registrar-General. These use the terms socio-economic groupings and Social-Classes. The latter bear a considerable resemblence to the idea of social classes suggested by Weber whereby the total structure of class and status relation-ships in society and the plural groupings that the complexity of their relationship must form merge into a clearly defined nexus related to the possibilities of movement between categories (Giddens, 1973, p. 48).

The classifications of the Census have been fivefold so far as Social-Classes are con-cerned. Broadly these have been:

1. Capitalists, manufacturers and professions
2. Small shopkeepers, lower professions etc.
3. Skilled workers
4. Semi-skilled labour
5. Unskilled labour.

They are obviously ranked in very general terms, but clearly overlap to a great degree. The quintet of Social-Classes today tends to form a somewhat normal distribution, with a modal class in Social-Class III. For example in 1966, the distribution of the econ-omically active population was as follows: I – 2.9 per cent; II – 14.6 per cent; III – 49.1 per cent; IV – 22.3 per cent; and V – 8.0 per cent (with 3.0 per cent unclassified) (Social Trends, 1972, p. 71).

In this study, the practice followed by Armstrong (1966, pp. 272–3) of using the Regis-trar-General's classifications has been adopted, but one major modification has been introduced by splitting Social-Class III into manual and non-manual occupations. In modern terms (for instance in 1966), this would divide Social-Class III on a 50/50 basis, but in the nineteenth century, it is likely that the balance was more in favour of manual occupations. This takes one on to considering all the many technical problems associated with adopting any occupational classification. These are of course especially severe in any study which covers a long period of time.

Armstrong (1972) has discussed the problems of occupational classifications in the mid-nineteenth century in some detail, though one is always finally left in these circum-stances with the excuse that there is no real alternative. Many of the problems – the difficulty of how people describe their jobs, the manner in which descriptions of such jobs change through time, the width of incomes and status that any specific description may entail – have been discussed elsewhere.[3] But there is one particular question that does need a final mention, and this is the relationship of any occupational classification to the structure of housing.

It is inevitable in such circumstances that one's arguments will contain a strong element of circularity. If one is forced to use any surrogate for the social structure which is as best 'fuzzy', at worst just unreliable, it is attractive to reverse one's opinions to start argu-ing that it is the housing structure that is the real picture of social structure and that whatever occupational groups fit any particular type or value of houses thereby repre-sents some significant dimension of the social structure. The truth is that unfortunately, like it or not, relationships, especially over a long period of time, are bound to be hazy, and one has to accept that matching these two elements must remain an imperfect exercise.

LEICESTER DIRECTORIES

The first urban gazetteers for the city of Leicester appeared in the 1780s as supplements to contemporary guides to the city. They primarily gave listings of the principal tradesmen. They were published irregularly into the nineteenth century, and in 1802 and 1815, Fowler's, a local publishing firm, brought out the first reasonably comprehensive listings of the more 'important' members of the community.

Strangely enough, the 1815 directory was the last for some fifty years to deal specifically with the city. In the 1820s, there grew up a more organised system of directory coverage which concentrated on producing 'county' directories. Two series covered Leicestershire – Kelly's and White's (which was based in Sheffield and concentrated on the East Midlands). Neither series appeared entirely regularly, though there was usually about a ten-year interval between editions. Coverage of Leicester itself was somewhat haphazard. Commercial information predominated, and sometimes Leicester received little more attention than small towns like Loughborough or Hinckley.

As cities grew, personal knowledge gave way to institutional sources of information. Advertising became important, and specifically urban directories began to appear in many larger towns. These were still financed by subscription, but this now meant advertising revenue. Usually, such directories had three or four sections – a street directory, in which occupiers were arranged geographically; a general index of residents represented; a commercial index of business firms; and sometimes a court directory listing the most important inhabitants. The first directory of this kind in Leicester appeared in 1870 with the Leicester Trade Protection Society directory edited by Christopher Norton Wright.[1] Further editions of this first modern-style directory for the city were issued in 1875 and 1878, but from 1880 onwards, the directories appeared under Wright's own name. Wright's directory of Leicester appeared continuously from 1880 to 1920. Usually, they came out biennially, with two three-year gaps from 1906 to 1909 and from 1911 to 1914, and a final six-year gap enforced by the First World War. The directories also covered the county area from the 1890s onwards, but always concentrated on the city.

Though Wright's activities did spread further afield, the firm remained concentrated in the East Midlands – Leicester, Nottingham and Sheffield. In 1902, the national firm of Kelly's began to publish special directories for the city itself to accompany the county directories that they had been publishing for some years. Kelly's Directories of Leicester appeared every three years between 1902 and 1914, but they were usually inferior to Wright's. Coverage varied far more and local areas appeared and disappeared from the street directory sections.

After 1920, Wright's went out of business and Kelly's took over much of their old records. In the 1920s, an improved version of Kelly's directory appeared, and at the end of the decade, Kelly's made great efforts to improve their directories in many medium-sized cities like Leicester. Council estates were included and the overall proportion of householders represented rose substantially. The directory for 1938 was probably the best record of this type ever produced for the city, and coverage might have improved still further had the war not intervened.

Post-war directories in Britain have not been very useful sources of information.

Occupational material has disappeared, and the directories have sometimes become nothing more than a partial and not always accurate summary of electoral rolls. Street and commercial sections remain and although coverage may reach 80 per cent in many medium-sized cities, certain groups like flat-dwellers are grossly underrepresented.[2]

ELECTORAL ROLLS

The Electoral Rolls upon which the analysis of movement in the city of 1870 has been based are very complex documents. Electoral registration was introduced by the Reform Act of 1832, and the power to draw up and administer the Rolls was placed in the hands of the Overseers of the Poor, who were the local authorities responsible for raising the Poor Rate. After 1835, with the passing of the 1834 Poor Law Amendment Act, these Overseers became Local Boards of Guardians.

The franchise of the 1832 Act was very restricted. In borough constituencies, a mass of 'ancient rights', local qualifications, was replaced by a qualification based on the occupation of rated property worth £10 per annum. Reference to the section on property rates and rents makes it clear how few people qualified under this franchise.[1]

At about the same time, however, the Municipal Corporations Reform Act of 1835 apparently created a much more democratic franchise for local elections, based on the vote for all adult male ratepayers who had been resident in the area for three years. Though this would seem to have been likely to produce a very great increase in the number of electors, in practice growth was severely curtailed by the spread of a common device known as 'compounding'. This occurred in many large towns (including Leicester) and involved the owners of property paying the rates for all their properties to the local authority in return for an overall rebate (known as a 'composition fee'). Consequently, many householders were not entered on the Rating Rolls. They simply paid their rates as part of their rent, and their names were never entered on the Rolls. In 1867, only 18 per cent of all householders in Leicester were on the Borough Electoral Roll.

This system was changed by the 1867 Reform Act. Two clauses are significant. Clause IV enfranchised all adult male householders in borough constituencies who had been living in the area for a year, whilst Clause VII (the 'Hodgkinson amendment') made 'compounding' illegal by compelling the compilers of the Rating Rolls to enter all the eligible occupiers of rated property rather than merely their landlords.[2] The act also introduced enfranchisement for all lodgers occupying property worth £10 per annum, but in practice this had little affect on the total numbers of electors.[3]

The 1867 Act raised the number of electors in Leicester from 3500 to 13,000. Such an increase of about three and a half times was above average but not exceptional. Leicester did have a fairly high ratio of electors to the total population (about 1:7.5). The registration process after 1867 became even more complicated because there were now a number of grounds under which objections could be lodged to any potential voter. The party organisations obtained a considerable role in trying to achieve the maximum registration of their own supporters and the minimum for their opponents.

After 1867, Electoral Rolls were supposed to be published on the 31 July in each year (though they usually did not become operational till the end of the year in question). Thus, to be qualified for entry on the Roll for 1871, a man would have had to have been entered by the 31 July 1870, and been resident and a ratepayer since the 31 July 1869. Thus the 1871 Roll would not have contained the names of persons who either had only entered in the last year or, for example, had married in the last year. Some persons were excluded for the reasons of objection and some for apathy and general disinterest. Some, too, were entered twice, though this appears to have been much more common in rural constituencies.[4]

NOTES

Chapter 2

1. In 1970, 70 per cent of all firms in the building industry employed seven workers or less, and only 210 out of a total of 73,000 firms employed more than 600 workers, though these 200 firms accounted for a third of the construction industry's output by value, and their importance has been increasing steadily since 1945 (Housing & Construction Statistics, 1972, pp. 63–4).

2. As with most topics, the Census has increased its interest in housing over the years. The 1911 Census was the first to take any great interest in housing matters, but tenure was not included till 1961.

3. The Central Housing Advisory Committee's report (CHAC, 1969), recommended that local authorities undertake local housing surveys as a basis for their policies. Surveys are also now frequently undertaken prior to the commencement of work in General Improvement Areas.

4. Since 1945, Britain has seen the creation of a number of 'New Towns', but even in these changes will occur (e.g. Heraud, 1968).

5. Conzen's work has been especially influential in encouraging work by historians, geographers and archaeologists on town plans (Lobel, 1969).

6. The manner in which immigrant groups have progressively changed their residential location within the city as they have moved up the social scale is a topic which is central to studies of American cities (see Rodwin, 1961; Glazer and Moynihan, 1962; Ward, 1971).

7. Schnore and Duncan have emphasised the theoretical links between human ecology and Durkheim's picture of urban, industrial society (Schnore & Duncan, 1959).

8. For the links between geography and town planning, see Freeman (1958). Not everyone has seen geographers' activities in planning as an unmixed blessing (e.g. Eversley, 1973).

9. Von Thunen's original work was republished in 1966 (Hall, 1966).

10. A 'bid-rent' is simply the price which any consumer can afford to go to for the purchase of a particular parcel of land. It is assumed that whoever can bid the most will be the one to occupy that land.

11. It is usual in such models to assume that accessibility has a monocentric distribution.

12. There are severe difficulties facing such a link (see Wilson, 1974, pp. 191–9).

13. There is nothing magical about 100 years. It is probable that this simply represents a combination of the rates of growth in cities in the late nineteenth century and present slum clearance efforts.

14. 'Site value' is the situation when a property has no other value than the land upon which it stands. It is by no means solely a theoretical concept since it is important in compensation payable in redevelopment schemes.

15. For geographical studies of the role of improvement policies see Hamnett (1973, pp. 252–60).

16. The 'community' and the 'neighbourhood' are by no means the same idea, and some sociologists have been very critical of the manner in which planners have interpreted the social consequences of physical developments (e.g. Willmott & Young, 1957, p. 198; and Dennis, 1958, pp. 191–206).

17. Strictly speaking, external economies are benefits and losses are external diseconomies. Externalities refer to both types.

18. It has been commonly observed that such threats may well act as a trigger to the formation of such action groups.

19. The results of the 1971 Census, which are now beginning to appear, do not indicate any great differences.

20. It is important to realise that the formation of households is not independent of the workings of the housing market. Couples may postpone their marriages if accommodation is not available.

21. This avoids the very difficult question as to what exactly is a 'household'. For an examination of the way in which the Census treats this problem see the General Explanatory Notes to the 1971 Census, County Volumes.

22. This description of the life-cycle is taken from Donnison (1967, pp. 215–16).

23. It is often the case in such circumstances that falling values can prevent the original population moving without severe financial loss (e.g. Smolensky, Becker & Molotch, 1968, pp. 419–30).

24. Appendix A discusses the available sources for intra-urban migration and appendixes D and E illustrate the way in which directories and Electoral Rolls have been used in this study.

25. For an introduction to Leicester, one cannot do better than the Handbook for the Annual Conference of the British Association for the Advancement of Science in 1972 (Pye, 1972).

26. Appendix A illustrates the present boundaries of Leicester and the divisions used in this study.

27. The one type of development not well represented in Leicester is the tenement, but in this Leicester is similar to most English cities outside London. For differences see Scotland and Tyneside (Cullingworth, 1967; Davies, 1972).

28. Neither of these points is completely true. There are industrial areas in Leicester today (e.g. in North Evington and along the north–south axis of the lines of the River Soar and the main railway line. Similarly, it will be seen later that physiographic factors did have some influence on the spatial evolution of the city in the middle of the nineteenth century.

Chapter 3

1. There are three classic works on the history of the industry: G. Henson, *The Civil, Political & Mechanical History of the Framework-Knitters of Europe & North America* (1831) and W. Felkin, *A History of the Machine Wrought Hosiery & Lace Manufacturers* (1867), both of which have been recently reprinted, and F. A. Wells, *The British Hosiery Trade* (London, 1935). 'Hosiery' rightly implies a concentration on the manufacture of stockings, and hosiery workers were often known as 'stockingers', but a wider range of products could be made on the hand-knitting frame invented by William Lea of Calverton at the end of the sixteenth century.

2. The highpoint of the Chartist disturbances was the so-called 'Battle of Mowmacre Hill' in August 1842.

3. There are two, indispensable, contemporary sources for the social condition of Leicester in this period. The state of the hosiery industry was such that it provoked a Royal Commission, which reported in 1845 and collected an enormous amount of evidence. The second source is the Annual Reports of the Town Missionary, Joseph Dare, published between 1845 and 1880.

4. Appendix A shows the administrative structure of city boundaries in this period.

5. Apart from the obvious difference of scale, London's pre-industrial size allowed a greater opportunity for the sub-division of older properties, and therefore greatly increased the importance of multi-occupancy in the housing market.

6. An excellent guide to the city in this period is provided by S. Watts, *A Walk Through Leicester* (1804) (Simmons, 1967).

7. The Annual Reports of the Medical Officer of Health are one of the best sources for the social situation in the city in the late nineteenth century, as well as for purely public health concerns.

8. The architect and main driving force behind this measure was Edwin Chadwick. The provision of the act to set up a Central Board of Health was in fact abolished in 1858, but the Act was important in stimulating a lot of local legislation over the next twenty years.

9. Ashworth quotes an interesting example from Leicester itself of the manner in which pre-1848 authorities found great opposition from their ratepayers to incurring expenditure on public health improvements.

10. Leicester Corporation introduced a system for the submission of plans for new building in 1849.

11. Enforcement in the 1850s in Leicester was lax mainly because penalties for offences were so nugatory. Cases of new dwellings 'falling down' soon after construction were reported in the Council Minutes.

12. Much of this section on the system of development in Leicester is based on the work of Graham Potts whose help and advice in these matters is gratefully acknowledged.

13. This is an indication of just how little the structure of the building industry has changed over the past century.

14. These two 'rivals' amalgamated in 1974.

15. There is a literally voluminous amount of material on the two Leicester building societies by way of records held in the Archives Department of the Leicestershire Records Office, and this represents the conclusions of only the most cursory analysis.

16. The information on the Leicester Freehold Land Society came from an analysis of the Society's Annual Reports in the Minute Books held in the County Records Office. For more information on the Leicester society see Pritchard (1972, pp. 71–3); and for a Birmingham example see Chapman & Bartlett (1971).

17. The basic qualification in borough constituencies after 1832 was the occupation of a house worth more than £10 per annum. Leicester was one of those boroughs where 'ancient rights' before 1832 had actually been a far more liberal franchise and it has been said that the borough's electorate was actually reduced by 7000 voters as a result of reform (Seymour, 1915, p. 84).

18. See Pritchard (1972, p. 73).

19. The nineteenth-century rating system is not an easy subject to comprehend. The introduction to the Royal Commission on the Ownership of Land (HMSO, 1871–3) is useful but The Institute of Municipal Treasurers, *The Rating of Dwellings: A History & General Survey* (London, 1962) is the best introduction.

20. Here, it is assumed that rateable value is a close approximation of rental value.

21. The 1855 Valuation was controlled by the 1836 Parish Assessments Act, and uses the concept of gross value as a measure of the supposed rental which a property might fetch on the market. It is known that before the Union Committee Assessment Act of 1862, standards of valuation were not high, but the range of values should reflect the general distribution of housing conditions and costs in the city.

22. In the nineteenth century, the differences between gross and net income were minimal for most people.

206

23. The two best candidates for the centre of Leicester are the Market Square or the Clock Tower at the bottom of Humberstone Gate, which is the 'symbol' of the city.

24. Taylor quotes evidence in Liverpool of densities of 1700 per acre (Taylor, 1970, pp. 186–97) and in very small areas in Leicester such as the Sandacre Street district, densities of over 1000 per acre seem possible.

25. Actually, only St Margaret's Parish is missing from the 1870 Valuations, but unfortunately that parish made up 60 per cent of the city's dwelling stock. In the other parishes, it was relatively easy to relate property values.

26. Many of these areas built in the 1860s were still inhabited in the middle and late 1960s when much of the fieldwork for this study was conducted.

27. Between 1851 and 1871, this parish's population fell from 2863 to 2507.

28. Under the franchise in operation, any property owner could be enfranchised and those owning non-residential property in the city had a vote. In local elections, this situation survived until 1969 when the 'business vote' was abolished.

29. This public ownership accounts for the high proportion of institutional uses in this area, especially Victoria Park and the University (originally the Lunatic Asylum!).

30. The association of poor Irish labourers with the worst quality housing in the city is one that especially struck Joseph Dare, for whom the Irish seem to have been something of a *bête noire*.

31. For the Reform Act of 1867 see Seymour (1915, pp. 258–311) and appendix E.

32. The lodger vote was in fact a damp squib; Seymour quotes that in 1868, only ten boroughs had more than 100 lodger voters (Seymour, 1915, p. 284): Leicester was no exception, it had five.

33. No detailed estimate for the number of women householders is possible for 1871, but in 1861, 15 per cent is a reasonable estimate according to the Enumerators' Returns (cf. Pritchard, 1972, p. 284).

34. Appendix E illustrates just how complicated the process of registration could be.

35. Leicester was divided into two Registration Districts, East and West, which when combined equalled the area of the Borough.

36. The Rolls are organised by wards and polling districts and alphabetically within the districts (of which there were twelve), this makes the process easier than would be so with the Rating Valuations.

37. In the 20–25 age-group, for example, only 40 per cent of men were married, and it is reasonable to assume that very few of this single majority had their own households.

38. Appendix E shows how the residence qualification imposes a time problem on the entry of new households on to the Electoral Roll.

39. This is a relative statement. What perhaps should be more accurately said is that there was no shortage of dwellings for those who felt they could afford a separate establishment.

40. There are no local figures for Leicester, but evidence later in the century from the MoH's Annual Reports suggests that this approximation holds good for this period. In 1901, the national average age of marriage for males was 27.2, compared to 24.6 in 1971 (*Social Trends*, 1972, p. 78).

41. In 1901, remarriages by men probably accounted for about 7.5 per cent of all marriages.

42. The role of the workhouse in the nineteenth-century housing market as a last resort is difficult to quantify. 35 per cent of Leicester's workhouse population in 1871 were over 60 (i.e. about 4 per cent of the total population over 60).

43. In all probability, the Electoral Rolls present the more stable elements in the population. They certainly do not include the army of travellers of all kinds who passed through the Common Lodging Houses, and it is probable that there were also a number of families, with their own dwellings, who spent only a short period of time in the city.

44. In this area, it is possible that the number of householders left off the Rolls because of omission, error etc. forms a greater proportion than in other parts of the city.

45. Migrants going to these areas were traceable through the directory.

46. For a description of this area which suggests some possible reasons for its aberrance, see Pritchard (1972, pp. 303), where it is suggested on the basis of 1861 evidence that this district may have been one which attracted fewer immigrants than other areas of similar housing structure in the city.

Chapter 4

1. Leicester benefited from a further burst of immigration in the 1870s and 1880s as a result of the Agricultural Depression, which especially affected the 'high farming' areas of the east of the country (Friedlander & Roshier, 1966, pp. 239–78; Saville, 1957).

2. This survey remains a central canon of Leicester folklore and even today no visitor can go for long without being reminded of it.

3. From the middle 1860s onwards, the small private, local bank gave way to limited liability companies which were able to lend capital on a larger scale but which were thereby more likely to demand the sort of security which this more formal pattern of development could provide (e.g. Ensor, 1936, p. 114).

4. As in 1870, the only parish missing is St Margaret's; furthermore the areas brought into the Borough by the extension of the early 1890s, Knighton, Aylestone and Humberstone are also available.

5. The 1911 Census was the first to look into housing matters in any great detail, and its introductory commentary to the state of housing in Britain is a most illuminating picture of the results of the late-Victorian development system.

6. The importance of the ability of the bulk of the population to afford the standards of dwelling demanded by public health legislation in the late nineteenth century has been stressed by a number of authors (see e.g. McKie, 1971). For general studies of the late-Victorian housing market see Tarn (1961); Hole (1965).

7. A 'building' in 1911 was any dwelling or group of dwellings with its own individual access to street, public landing or staircase.

8. A 'block of flats' in 1911 would have included only purpose-built properties, plus the sort of tenements common in Scotland and some parts of the North-East of England.

9. For the ward boundaries in this period see appendix B.

10. As a general indication of this inflation, all items in the cost of living index rose by 25 per cent between July 1914 and July 1915. Food prices rose 32 per cent (Dept of Employment, 1971, Table 89, p. 166).

11. The post-war period saw an explosion in housing costs. The index for rent + rates costs rose by 44 per cent between January 1919 and January 1924, although all costs fell by about a quarter in the same period. From 1924 to 1939, housing costs rose by only about a further 8 per cent (Dept of Employment, 1971, Table 89, pp. 167–8).

12. To give some indication of the results, the amount lent by building societies in a year went up from under £10 million in 1910 to nearly £90 million in 1930 and £140 million in 1938, when there were nearly a quarter of a million mortgages and an average loan was about £600 (Rollett, 1972, p. 317).

13. At a rough approximation, there were over 200 houses in this area in multi-occupancy in 1938; see Pritchard (1972, p. 106).

14. For directories as a source see appendix D.

15. The sample was drawn by the simple method of selecting every twentieth name from the private index sections of the directories.

16. Appendix C deals with this specially thorny question in more detail.

17. Geographers have tended to emphasise the role of public transport and the development of accessibility in determining the pattern of urban growth; something that has received less attention has been the layout of such public infrastructure lines and the role played by legislation in giving rights of connection to such lines if development is within certain limits.

18. 'Fever' is a term much used by Victorian public health officials. It usually refers to some sort of typhoid or paratyphoid disease associated with poor standards of public hygiene.

19. For the development of the tramway system in Leicester which especially emphasises its social role and the manner in which the tramways were influenced by political organisation see Murphy (1974).

20. Part of this district was demolished in 1972 to make way for a new road scheme.

21. One of Leicester's few architecturally distinguished buildings, Ernest Gimson's 'White House' in North Avenue is in this area.

22. All modern, privately built flats are 'luxury'!

23. The previous chapter discusses the problem of the Freehold Land Societies.

24. The occupational classification used is described and discussed in appendix C. Basically, it is the normal, fivefold classification used by the Registrar-General, with the exception that Social-Class III has been split into non-manual (IIIa) and manual (IIIb) sections.

25. There could have been only a very limited number of divorced people in any of these areas before the Second World War. For example before 1914, *nationally*, there were much fewer than a thousand divorces a year. For the whole problem see McGregor (1957).

26. H. J. Dyos' as yet unpublished study of post-1871 Camberwell seems to indicate something similar (see Dyos & Reeder, 1973, p. 375).

27. Obviously, compiling directories is a continuous process, and probably the best one can hope is that between individual areas a similar time-lag existed. It does provide a problem because it may be that the real time intervals referred to vary between areas.

28. Limited liability became practically possible on a large scale as a result of Acts of 1856 and 1862 (Thompson, 1950, p. 140).

29. Though one can make attempts to build income data into such a classification at a national scale, at the local level in a study of this kind it is virtually impossible.

30. The three categories are defined by whether persons were living at the same address in the next directory, were living at another address in the Leicester area recorded in the directory, or were completely lost. The directories in practice, especially for the upper sample, do cover areas outside the city boundary.

31. This is the sample used earlier to ascertain the structure and reliability of the directories' coverage.

32. If anything, the tendency for older people to leave the housing market to live with children and so forth may have produced a real distribution which was younger.

33. It is also highly likely that these changes fit into the altering pattern of the relationship of mobility and social class. Differentials in death rates between working- and middle-class householders have probably shrunk in the last century and this would contribute to the greater fall in mobility the lower down the social scale one goes.

34. 'Entry-groups' have been used rather than the more usual demographic term 'cohorts' because of the inexactitude of their age-structure.

35. This argues that age-structure has always been more important for the upper sample and may tie in with the already observed fact that the upper sample areas had a higher proportion of widows as householders in the nineteenth century.

36. This is a peripheral, but extremely interesting comment. Social mobility has always been a central concern in American urban studies, even if the intention has been to question or revise its importance (see Thernstrom, 1964). (It is of course a view fundamental to the ecological model of the city.) Social mobility in Britain is a subject which has not received anything like the same attention (though see Glass, 1954), but is now receiving more attention (see MacDonald & Ridge, 1972, pp. 129–47; Ridge, 1974). In this country, especially in a historical context, class and class formation remain major concerns, particularly for Marxists (see Foster, 1974).

37. For a discussion of this see Katz (1970), pp. 209–44.

38. They are considerably less for the directories than for the Electoral Rolls because of the usual combination of names and occupational data.

39. This view is of course reinforced by the directories' better coverage in the inter-war period, and the property that they adopted then of recording 'community leaders' even if they were living some distance away from the built-up area of the city itself.

40. This is probably more important in a provincial city like Leicester which primarily has a lot of branch or regional offices than it might be in London.

41. In stating this, one must exclude all those vast numbers of movements made during the war which were of a temporary nature, including evacuation, war-work and service in the armed forces. Indeed, given the enormous scale of such movements, the extraordinarily limited effects of the war on residential location patterns are quite remarkable.

42. In using the turnover rate to approximate to variations in mobility in the city, it must be realised that this will include such factors as those changes brought about by the deaths of householders, though it may be hoped that these are relatively constant.

43. The annual figures were taken from the Medical Officer of Health's Reports and were converted to the periods covered by the directories.

44. For roughly comparable figures for Great Britain see Hole and Pountney (1971, p. 30).

45. It is here assumed that a shortage of dwellings must have produced higher real costs in housing and a surplus of property lower. Unfortunately, detailed evidence for Leicester is not available.

46 The advantage of the non-parametric measure is that it avoids some of the more obvious difficulties associated with the comparison of time series, and will absorb a greater degree of error in the data.

Chapter 5

1. This meant that some cheaper properties had been in control since 1915.

2. Cullingworth's suggestion applies to the national situation and whether this was true in Leicester is uncertain. It is probable that in view of the city's attraction for migrants in the 1930s (and the evidence of shortage presented by the growth of multi-occupancy), it was not. The city did not suffer from great air-raid damage but there was the general deterioration of the stock.

3. No one ever defined who were the 'working classes'.

4. The main changes have been concerned with the problem of development value. The 1947 Act nationalised development value, but this system was progressively dismantled by Acts of 1953, 1954 and 1959. The Land Commission of 1964 marked another attempt to deal with the problem but was abolished in 1970. In early 1975 the government, through the Community Land Bill, introduced another attempt to solve this problem through the municipalisation of development land. Their proposals are set out in the White Paper, *Land* (Cmnd 5730) published in September 1974. (For an introduction see Cullingworth, 1970.)

5. See Hall (1973) for an assessment of the relationship of the statutory planning system to post-war patterns of urban development (especially chapter ix which deals with Leicester).

6. In 1971 and 1972, private building accounted for 200,000 new houses out of about 350,000. The most useful source of statistics for housing and construction are the Dept of the Environment's quarterly *Housing & Construction Statistics*.

7. In 1972, £4,310 millions were lent for house purchase, £3,649 millions by building societies, £325 millions by banks, £191 millions by local authorities and £145 millions by insurance companies (*Housing & Construction Statistics*, table 35).

8. In 1972, the average incomes of persons taking out mortages from building societies was £2,474 (*Housing & Construction Statistics*, table 38) and this compares with an average household income in the same year of £2,228 (*Social Trends*, 1973, p. 102).

9. The Housing Finance Act, 1972, altered this system but is now to be abolished and replaced by a new system of local authority housebuilding finance.

10. In the year 1972–3, central government subsidies on housing totalled £294 millions in a total public expenditure of £1,655 millions.

11. The Parker Morris standards (*Homes for Today & Tomorrow*, HMSO, 1961) were a long time in being adopted because of constraints of economy, and its recommendations were not implemented till 1969.

12. What is 'unfit' remains somewhat arbitary; the 1957 Housing Act adopted eight areas in which dwellings could be unfit and much depends on the interpretation given by Public Health Inspectors (Cullingworth, 1966, pp. 177–81). The 1971 House Condition Survey suggested there were 1,244,000 'unfit' dwellings (7.3 per cent of total) and another 2,866,000 (16.8 per cent) lacking a basic amenity (Smith, 1974, p. 11). In 1971, 34 per cent of households in Leicester lacked the exclusive use of such basic amenities.

13. Improvement grants have existed since the 1949 Housing Act, but their role in the market was considerably increased in 1969 with the introduction of the General Improvement Area, which combined grants with environmental improvements; between 1969 and 1972, nearly 600 General Improvement Areas have been declared, containing over 180,000 dwellings (*Social Trends*, 1973, p. 161). For improvement in Leicester see Anson & Shelton, 1971, pp. 229–30, which makes some very good points about the role of inner areas in the urban social structure.

14. In 1972, housing associations accounted for about 2 per cent of all completions of dwellings. For this section of the market see *Housing Associations: A Working Paper of the Central Housing Advisory Committee* (HMSO, 1971).

15. A 'fair rent' is assessed by reference to the character, location and amenities and state of repair, but disregarding the value due to any local shortage of similar accommodation.

16. In the early 1970s, slum clearance in England and Wales has been running at between 65,000 and 70,000 houses per annum (*Housing & Construction Statistics*, table 32).

17. The 1974 Rent Act extended the same rights concerning security of tenure to furnished tenants.

18. This is the same sort of methodology as used in Robson, (1966); Gordon (1971).

19. The plan was the result of the Survey to the City Development Plan (Leicester City Council, 1952).

20. This has been part of a general clearance programme which by 1974 had cleared a vast area all along the east side of Belgrave Road, stretching well north of the old Great Northern Railway line.

21. Nationally, slum clearance and other losses of residential property are at present accounting for a net decline of about 120,000 dwellings in this tenure every year. At this rate, the tenure has about twenty years 'life'.

22. 'Sharing' is now an extremely complex concept in the Census, which is a combination of structural and family relationships. For an explanation see the *General Explanatory Notes: County Volumes, 1971 Census.*

23. Some groups such as students and the first wave of Commonwealth immigrants, concentrated in the younger male age-groups, probably actually preferred to share.

24. King (1969, pp. 166–84) gives an excellent introduction to the technique, slanted towards its use in geography.

25. Berry (1961; 1966) shows some of the earliest examples of this form of the technique.

26. The 1971 Census Ward Library material has only just become available in 1974.

27. It will later be seen that the detailed year to be studied will be 1963–4, and this year lies between the two Censuses.

28. For each of the seven components, the 'significant' loadings are defined as $+$ or $-$ 0.5, and the total 'rotated' variance is also shown.

29. Each map shows the pattern of scores about a mean of zero, and with a variance of unity.

30. The skew of this component towards such poorer areas is exaggerated by the exclusion from the analysis of ex-city suburban areas such as Oadby and Wigston, which could be expected to have largely positive scores (see Lewis, 1972, pp. 455–80 for a wider analysis on this basis).

31. Asian households were 'recognised' by their distinctive names.

32. The classification procedure was undertaken through the Clustan Ia Classification Package drawn up by D. Wishart of the University of St Andrews.

33. The two inter-war council estates not in Group I are Northfields, which is really a mixture of pre- and post-1945 developments, and the 'youngest' of the earlier areas, that part of Braunstone built in the middle and late 1930s.

34. The 1966 Census contains Five-Year Mobility tables, the 1961 only One-Year tables. The Five-Year have a larger sample, and since all mobility figures are on a 10 per cent sample, they have an advantage.

35. See for example Donnison (1967, pp. 197–214).

36. The 1969 Representation of the People Act lowered this age to 18.

37. There are lots of possibilities and every election brings forth its oddities. It is known that error tends to be concentrated in furnished letting with its transitory population.

38. Adult suffrage means that the real number of household units in Leicester was only about 85,000.

39. The areas are shown in appendix A.

40. The logic here is that the Electoral Roll entries are fuller, but the indexing system of the Telephone Directory makes it easier to use in the initial search procedure.

41. The difference between a hereditament and a household is basically the difference between a structural and a usage definition of property holding.

42. The peak Electoral Roll population was reached in 1955, and between 1955 and 1964 there was a total loss of some 14,000 voters from the city.

43. The distinction between household and non-household moves is a difficult one to examine. Here, a household is any person or group of persons who move over a similar path.

44. In 1963–4, 42 per cent of women marrying were under 21.

212

45. In 1963–4, it was still usual, with the voting age at 21, for students to remain on their home register, even if they passed the age of qualification, and it is generally likely that many single person moves in the city were of people under 21.

46. This is an underestimate of the total of number of persons reaching 21 in this year as far as it can be calculated from the Census, and may be because there is some doubt whether the 'Y' voter procedure, which put people reaching the age of enfranchisement during the period of the roll on the list, was fully understood prior to its simplification in the 1969 Act.

47. It is hoped that this is a rough proxy which eliminates those persons who were merely sharing with in-laws. In 1971, there were 4200 households in shared accommodation in the city.

48. This excludes those council properties outside the city boundary, e.g. in Glenfields, Glen Parva and Scraptoft.

49. There are some pockets of post-1920, owner-occupied dwellings which are rated at less than £70 per annum. A substantial district is in Belgrave, where subsidised housing built in the early 1920s was originally let but a great many were sold to sitting tenants. These properties are rated at between £55 and £65 per annum at pre-1972 valuations.

50. Some of these immigrants, of course, were new Asian households who were concentrated in this category of housing.

51. It is extraordinarily difficult to define what is a household in this sector from the Electoral Roll material, so this must be an approximation.

52. For an investigation into the relationship of housing and the mobility of households see Johnson, Salt & Wood (1974); Bell (1969) emphasises the role of the middle classes.

Chapter 6

1. This diagram is of quite complex construction. The natural change and migration figures are derived from the Registrar-General's Mid-Year Estimates. The figures for 1931–41 and 1941–51 are calculated from the Medical Officer of Health's estimates. Also, by 1961–71, even the combination of Leicester, Oadby and Wigston no longer represents the real growth area of the city, because population expansion by this period was concentrated in the surrounding rural districts which are not here represented.

2. This has been most marked since 1911, when the major fall in mobility has occurred. Between 1911 and 1971, the population of the City of Leicester increased from 227,000 to 284,000 (+ 25 per cent); by contrast, the number of households increased from 51,500 to 95,500 (+ 85 per cent).

3. A 'headship rate' is simply the proportion of persons in the population who head households, and is often considered in terms of sub-groups such as married men, divorced persons etc.

Appendix A

1. The best introduction is Welch (1971).

2. Sweden, which has one of the best population monitoring systems in Europe, has produced much original research on migration e.g. the work of Torsten Hagerstrand.

3. The Registrar-General does publish annual mid year estimates for the populations of local authorities which have estimates of net migration within them.

4. An exception is the recent survey carried out in West Yorkshire using the facility of the Electricity Boards' records.

5. *Papers on Migration and Mobility in North East England,* commissioned by the Ministry of Labour from J. W. House at the University of Newcastle.

Appendix B

1. The *Leicester Daily Post* of 20 October 1890 gives full details of this boundary extension. The ward revision was carried out under Local Government Board Order, No. 32,954 (26 March 1896). For the 1935 revision see the *Leicester Mercury* of 29 March 1933 and also that same paper's special supplement on 'Great Leicester' of 1 April 1935.

Appendix C

1. For introductions to this necessarily complex topic see Jackson (1968) and Beteille (1969). For an extremely useful and interesting recent interpretation of social class see Giddens (1973).

2. A fascinating view of just how significant such syntactical problems are, especially when considered historically, is given by Briggs (1960).

3. Apart from Armstrong, other worthwhile references especially include Moser & Hall (1954), which remains the basis for most of the official classifications and therefore for much academic research.

Appendix D

1. C. N. Wright lived for some years in West Street in the Southfields district of the city, and as such appears in the residents of Area 1 in the ten sample areas of chapter 4.

2. It is chastening to compare the character of British directories with the use made of the same source in the United States. In a historical context see particularly Thernstrom and Sennett (1970), especially the chapters by Thernstrom, Knights, and Knights and Schnore. See also Knights (1971).

Appendix E

1. The indispensable source for the nineteenth-century franchise is C. Seymour, *Electoral Reform in England & Wales* (Yale, 1915) reprinted 1970, with an introduction by M. Hurst. Seymour quotes Leicester as perhaps the best example of a place where the electorate actually fell as a result of the 1832 Act (he quotes a loss of 7000 voters, which must have meant a fall to only about 30 per cent (Seymour, 1915, p. 84)).

2. Seymour points out that this clause seems to have received great opprobrium from those it was meant to benefit, since the abolition of composition fees were used as an excuse to raise rents. In practice, the law lapsed in that composition was restored but it was tacitly agreed that tenants in such properties should be entered on the Rolls (Seymour, 1915, p. 354–6).

3. Leicester had five enfranchised lodgers in 1871.

4. Seymour quotes the plural franchises in county constituencies as a major cause (Seymour, 1915, p. 363). There is not great evidence that Leicester was especially afflicted by this problem.

BIBLIOGRAPHY

A. Primary sources

1. Directories

Fowler's Directory of Leicestershire (Leicester, 1815)
White's Directory of Leicestershire & Rutland (Sheffield, 1846; 1863)
Leicester Trade Protection Society, Directory of Leicester (Leicester, 1870; 1875 and 1878)
C. N. Wright's Directory of Leicester (Leicester, 1880; 1882; 1884; 1886; 1888; 1890; 1892; 1894; 1896; 1898; 1900; 1902; 1904; 1906; 1909; 1911; 1914; 1920)
Kelly's Directory of Leicestershire and Rutland (London, 1907; 1910; 1912; 1922; 1925; 1928; 1932; 1936)
Kelly's Directory of Leicester (Kingston-upon-Thames, 1938; 1941; 1947; 1951; 1954; 1957; 1960; 1963; 1966; 1969)

2. Electoral Rolls

Leicester Borough Burgess Lists, 1835–66
Leicester City Constituencies; Parliamentary Electoral Rolls, 1868–1964
Melton, Harborough & Bosworth County Constituencies; Electoral Rolls, 1963–64

3. Rating registers

Leicester Borough General Rate Assessment, 1854–1855 (5 vols.), held in Leicestershire County Records Office, Archives Dept
Leicester City Rate Revaluation, 1895–1896 (7 surviving volumes), held in Leicestershire County Records Office, Archives Dept
Leicester City Rate Revaluation, 1963–64 (14 vols.), held in City Rates Office

4. Census of Population

References to the decennial Census published volumes are so numerous that details have not been given in the text, but note should also be taken of the following unpublished sources of Census information that were used:
1861 Census – Enumerators' Books
1871 Census – Enumerators' Books
   Microfilm copies of both of these sources are held in the County Records Office
1961 Census – Ward Library, held in Leicester City Planning Office
1966 Census – Ward Library, supplied by the Registrar-General

5. Miscellaneous

Annual Reports of the City Medical Officer of Health (Leicester, from 1848)
Annual Reports of the Leicester Domestic Mission, ed. J. Dare (Leicester, 1845–80)
Minutes of the Leicester City Council (original copies of pre-1892 Minutes held in Leicestershire County Records Office. Post-1892 bound volumes in Leicester City Library)
Minutes of the Leicester Freehold Land Society (Leicester, 1848–92), held in County Records Office, Archives Dept
Annual Reports of the Leicester Permanent Building Society (Leicester, 1855–1914), held in County Records Office, Archives Dept

215

B. Secondary sources

ADAMS, J. S. (1969), 'Directional bias in intra-urban migration', *Econ. Geogr.*, **45**, 302–23

ADAMS, J. S. (1970). 'Residential structure of mid-western cities', *Ann. Assoc. Amer. Geogr.*, **60**, 37–62

ALDCROFT, D. H. & RICHARDSON, H. A. (1968). *Building in the British economy between the wars*, Univ. of Glasgow Soc. & Econ. Studies (London)

ALDCROFT, D. H. & RICHARDSON, H. A. (1969). *The British economy, 1870–1939* (London)

ALLAN, C. M. (1965). 'The genesis of British urban redevelopment, with special reference to Glasgow', *Econ. Hist. Rev.* 2nd Series, **18**, 598–613

ALONSO, W. (1964). *Location and land use* (Harvard)

ANDERSON, M. (1971). *Family structure in nineteenth century Lancashire* (Cambridge)

ANSON, D. J. & SHELTON, A. J. (1971). 'Social aspects of improvement', *J.R.T.P.I.*, **57**, 229–30

ARMSTRONG, W. A. (1966). 'Social structure from the early Census returns', in *An introduction to English historical demography*, ed. E. A. Wrigley, pp. 209–37 (Cambridge)

ARMSTRONG, W. A. (1972). 'The use of information about occupation', in *Nineteenth century society: Essays in the use of quantitative methods*, ed. E. A. Wrigley, pp. 191–310 (Cambridge)

ARMSTRONG, W. A. (1974). *Stability and change in an English county town: A social study of York, 1801–1881* (Cambridge)

ASHWORTH, W. (1954). *The genesis of British town planning* (London)

BARBOLET, R. H. (1969). 'Housing classes and the socio-ecological system', *Centre for Environmental Studies WP4*

BELL, C. R. (1969). *Middle-class families: Social and geographic mobility* (London)

BERRY, B. J. L. (1961). 'A method for deriving multi-factor uniform regions', *Przegl. Geogr.*, **33**, 263–82

BERRY, B. J. L. (1966). *Essays on commodity flows and the spatial structure of the Indian economy*, Univ. of Chicago Research Paper, 111

BESHERS, J. M. (1962). *Urban social structure* (New York)

BEST, G. (1971). *Mid-victorian Britain, 1851–1875* (London)

BEST, R. H. & COPPOCK, J. T. (1962). *The changing use of land in Britain* (London)

BETEILLE, A. (1969). *Social inequality* (Harmondsworth)

BOAL, F. W. (1969). 'Territoriality on the Shankill–Falls divide, Belfast', *Irish Geogr.*, **6**, 30–50

BOAL, F. W. (1972). 'The urban residential sub-community: A conflict interpretation', *Area*, **4**, 164–8

BOOTH, C. (1889). *Life and labour of the people*, vol. I (London)

BOTT, E. (1957). *Family and social network* (London)

BOWLEY, M. W. (1944). *Housing and the state* (London)

BOYCE, R. (1969). 'Residential mobility – its implications for urban spatial change', *Proc. Assoc. Amer. Geogr.*, **1**, 22–6

BRACEY, H. E. (1964). *Neighbours on new estates and subdivisions in England and U.S.A.* (London)

BRIGGS, A. (1960). 'The language of "class" in early nineteenth century England', in *Essays in Labour history*, ed. A. Briggs and J. Savile, pp. 43–73 (London)

BROWN, L. A. & MOORE, E. G. (1970). 'The intra-urban migration process: A perspective', *Geogr. Annaler*, **52B**, 1–13

BROWN, L. A. & LONGBRAKE, D. B. (1970). 'Migration flows in intra-urban space: Place-utility considerations', *Ann. Assoc. Amer. Geogr.*, **60**, 368–84

BROWN, L. A. & HOLMES, J. (1971). 'Search behaviour in an intra-urban migration context: A spatial perspective', *Env. and Planning*, **3**, 307–26

BURGESS, E. W. (1925). *The urban community* (Chicago)

216

BURGESS, E. W. (1928). 'Residential segregation in American cities, *Ann. Amer. Acad. Pol. & Soc. Sci.*, **140**, 105–15

BURGESS, E. W. & BOGUE, D. J. (1964). *Contributions to urban sociology* (Chicago)

BURGESS, E. W., PARK, R. E. & MCKENZIE, R. (1925). *The city* (Chicago)

BYATT, I. C. R., HOLMANS, E. A. & LAIDLER, D. E. W. (1972). *Income and the demand for housing: Some evidence for Great Britain*, Dept of the Environment, Economic and Statistical Notes, 1

CAIRNCROSS, A. K. (1949). 'Internal migration in Victorian England', *Manchester School*, **17**, 67–87

CAIRNCROSS, A. K. (1953). *Home and foreign investment* (London)

CARTER, H. (1968). *The town of Wales* (Cardiff)

CAVE, P. W. (1969). 'Occupancy, duration and the analysis of residential change', *Urban Studies*, **6**, 58–69

CHAC (Central Housing Advisory Committee) (1961). *Homes for today and tomorrow* (The Parker Morris Report) (HMSO)

CHAC (Central Housing Advisory Committee) (1967). *The needs of new communities* (HMSO)

CHAC (Central Housing Advisory Committee) (1969). *Council housing: Purposes, procedures and priorities* (HMSO)

CHAC (Central Housing Advisory Committee) (1971). *Housing associations* (HMSO)

CHADWICK, E. (1842). *Report on the sanitary condition of the labouring classes* (London)

CHAMBERLAIN, B. G. (1861). *Report on the administration of the Poor Law: Leicester Union* (Leicester)

CHAPMAN, D. (1955). *The home and social status* (London)

CHAPMAN, S. D. (1965). 'The transition to the factory system in the Midlands cotton spinning industry', *Econ. Hist. Rev.*, **18**, 526–43

CHAPMAN, S. D. (1969). 'Working class housing in Nottingham during the Industrial Revolution', *Trans. Thoroton Soc. Nottingham*, **48**, 69–92

CHAPMAN, S. D. (1971). *Working-class housing: a symposium* (London)

CHAPMAN, S. D. & BARTLETT, J. N. (1971). 'The contribution of building clubs and freehold land societies to working-class housing in Birmingham', in *Working-class housing: A symposium*, ed. S. D. Chapman, pp. 223–46 (London)

CHERRY, G. E. (1972). *Urban change and planning* (London)

CHERRY, G. E. (1974). *The evolution of British town planning* (London)

CHILTON, R. & POET, R. R. W. (1973). 'An entropy maximising approach to the recovery of detailed migration patterns from aggregate Census data', *Env. and Planning*, **5**, 147–56

CLARK, D. (1973). 'Urban linkage and regional structure in Wales: An analysis of change, 1958–68', *Trans. Inst. Brit. Geogr.*, **58**, 41–58

CLARK, W. A. V. (1969). 'Information flows and intra-urban migration: An empirical analysis', *Proc. Assoc. Amer. Geogr.*, **1**, 38–42

CLARK, W. A. V. (1970). 'Measurement and explanation in intra-urban residential mobility', *Tijdschrift v. Econ. Soc. Geogr.*, **61**, 130–54

CLEARY, E. J. (1965). *The building society movement* (London)

COLLISON, P. (1959). 'Occupation, education and housing in an English city', *Amer. J. Soc.*, **64**, 583–97

COLLISON, P. (1963). *The Cutteslowe walls* (London)

CONZEN, M. R. G. (1960). *Alnwick* (London)

CULLINGWORTH, J. B. (1963). *Housing in transition: A case study in the city of Lancaster* (London)

CULLINGWORTH, J. B. (1965). *English housing trends: A report on the Rowntree Trust housing study* (London)

CULLINGWORTH, J. B. (1966). *Housing and local government* (London)

CULLINGWORTH, J. B. (1967). *Scottish housing in 1965* (HMSO)

CULLINGWORTH, J. B. (1970). *Town and country planning in England and Wales* (London)

CULLINGWORTH, J. B. & ORR, S. C. (eds.) (1969). *Regional and Urban Studies* (London)

Bibliography

DAHRENDORF, R. (1959). *Class and class conflict in industrial society* (Stanford)
DARBY, H. C. (1943). 'The movement of population to and from Cambridgeshire between 1851 and 1861', *Geogr. J.* **101**, 118–25
DAVIES, J. G. (1972). *The evangelistic bureaucrat* (London)
DAVIES, J. G. & Taylor, J. (1970). 'Race, community and no conflict', *New Society*, 9 July 1970
DENNIS, N. (1958). 'The popularity of the neighbourhood community idea', *Soc. Review*, VI (2) (1958), reprinted in *Readings in urban sociology*, ed. R. E. Pahl pp. 74–94 (London, 1968)
DEPARTMENT OF EMPLOYMENT (1971). *British labour statistics: Historical abstract, 1886–1968* (HMSO)
DoE (Department of the Environment) (1972). *The estate outside the dwelling*, Design bulletin 25 (HMSO)
DEWEY, S. (1948). 'Peripheral expansion in Milwaukee county', *Amer. J. Soc.*, **53**, 166–89
DICKINSON, R. E. (1947). *City, region and regionalism* (London)
DONNISON, D. V. (1961). *Housing policy since the war* (London)
DONNISON, D. V. (1967). *The government of housing* (Harmondsworth)
DUNCAN, O. D., CUZZORT, R. P. & DUNCAN, B. (1961). *Statistical geography* (Glencoe, Ill.)
DUNCAN, T. L. C. (1971). *Measuring housing quality: A study of methods*, Occ. Paper 20, Centre for Urban & Regional Studies, Univ. of Birmingham
DUNN, M. C. & SWINDELL, K. (1972). 'Electoral registers and rural migration: A case study from Herefordshire', *Area*, **4**, 39–41
DURY, G. H. (1963). *The East Midlands* (London)
DYOS, H. J. (1961). *Victorian suburb* (Leicester)
DYOS, H. J. (1968). *The study of urban history* (London)
DYOS, H. J. & REEDER, D. (1973). 'Slums and suburbs', in *The Victorian city*, ed. H. J. Dyos and M. Woolf, pp. 359–86 (London)
EDWARDS, K. C. (1964). 'The growth of Nottingham', in *Nottingham and its region*, ed. K. C. Edwards pp. 1–30 (Nottingham)
ELLIS, E. H. (1948). *History in Leicester* (Leicester)
ENSOR, R. C. K. (1936). *England, 1870–1914* (Oxford)
EVANS, R. H. (1972). 'Leicester & Leicestershire, 1835–1971', in *Leicester and its Region*, ed. N. Pye, pp. 288–310 (Leicester)
EVERSLEY, D. E. C. (1973). *The planner in society* (London)
Felkin's history of the machine wrought hosiery and lace manufacturers (1967). Centennial edition, ed. S. D. Chapman (Nottingham)
FIELDING, A. J. (1969). 'Internal migration in England & Wales', *Centre for Environmental Studies WP4*
FORSTER, C. A. (1973). *Court housing in Kingston upon Hull: An example of cyclic processes in the morphological development of nineteenth century bye-law housing*, Univ. of Hull, Occ. Papers 19
FOSTER, J. (1968). 'Nineteenth century towns – a class dimension', in *The study of urban history*, ed. H. J. Dyos, pp. 281–300 (London)
FOSTER, J. (1974). *Class struggle and the industrial revolution* (London)
FRANCIS COMMITTEE (1971). *Report on the workings of the Rent Acts* (HMSO)
FRANKENBERG, R. (1966). *Communities in Britain* (Harmondsworth)
FREEMAN, T. W. (1958). *Geography and planning* (London)
FRIEDLANDER, D. & ROSHIER, R. J. (1966). 'A study of internal migration in England & Wales', *Pop. Studies* **19**, 239–79, **20**, 45–59
GANS, H. J. (1958). *The urban villagers* (Boston)
GERMANI, G. (1964). 'Migration and acculturation' in *UNESCO Handbook for Social Research in Urban Areas*, ed. P. Hauser, pp. 159–78 (UNESCO)
GHS (1973). *General Household Survey* (HMSO)
GIDDENS, A., (1973). *The class structure of the advanced societies* (London)

218

GITTUS, E. (1964–5). 'The structure of urban areas', *Town Planning Review*, **35**, 5–20.
GLASS, D. V. (1954). *Social mobility in Britain* (London)
GLASS, R. (1955). 'Urban sociology in Great Britain', *Current Sociology*, **4**, 5–21
GLASS, R. (1969). 'Housing in Camden', *Town Planning Review*, **39**, 15–40
GLAZER, N. & MOYNIHAN, D. P. (1962). *Beyond the melting pot* (New York)
GODDARD, J. (1970). 'Functional regions within the city centre: A study by factor analysis of taxi flows in Central London', *Trans. Inst. Brit. Geogr.*, **49**, 161–82
GOHEEN, P. (1969). 'The North American industrial city in the late nineteenth century: The case of Toronto', Dept of Geography Research Paper, Chicago
GOLDSCHREIDER, C. (1966). 'Differential residential mobility of the older population', *J. Gerentology*, **21**, (1), 103–8
GOLDSTEIN, S. & MAYER, K. (1965–6). 'The impact of migration on the socio-economic structure of city and suburbs', *Soc. & Sociologial Research*, **26**, 5–23
GORDON, G. (1971). 'Status areas in Edinburgh', Edinburgh University PhD Thesis
GRAY, P. G. & RUSSELL, R. (1962). *The housing situation in 1960: An enquiry covering England & Wales carried out for the Ministry of Housing and Local Government*, Social Survey, HMSO
GREBENIK, H. A. & SHANNON, D. (1943). *The population of Bristol, 1938* (London)
GREENBIE, B. B. (1969). 'New house or new neighbourhood? A survey of priorities among home owners in Madison, Wisconsin', *Land Econ.*, **45**, 234–56
GRIGSBY, W. G. (1963). *Housing markets and public policy* (Philadelphia)
HABBAKUK, H. J. (1962). 'Building and the British economy', *J. Econ. Hist.*, **22**, 198–230
HALL, P. G. (1966). *Von Thunen's isolated state* (London)
HALL, P. G. (1973). *The containment of urban England* (London)
HAMNETT, C. A. (1973). 'Improvement grants as an indicator of gentrification in Inner London', *Area*, **5**, 252–60
HARRISON, J. F. C. (1959). 'Chartism in Leicester', in *Chartist Studies*, ed. A. Briggs, pp. 99–146 (London)
HARTOPP, H. (1933). *Register of the freemen of Leicester*, 2 vols. (Leicester)
HARVEY, D. (1970). 'Social processes, spatial form and the redistribution of real income in an urban system', in *Regional forecasting*, ed. M. Chisholm, A. E. Frey & P. Haggett, pp. 270–300 (Bristol)
HARVEY, D. (1972). *Social justice and the city* (London)
HATT, P. & REISS, A. J. (1957). *Cities and society* (Glencoe, Ill.)
HAUGHTON, J. P. (1949). 'The social geography of Dublin', *Geogr. Review*, **39**, 257–77
HAWLEY, A. H. (1950). *Human ecology* (New York)
HEAD, P. (1961–2). 'Putting-out in the Leicester hosiery industry in the middle of the nineteenth century', *Trans. Leics. Archaeol. Soc.*, **37**, 44–59
HEAP, D. (1973). *An outline of planning law* (London)
HENSON, G. (1831). *The civil, political and mechanical history of the framework-knitters of Europe and North America* (Leicester)
HERAUD, B. J. (1968). 'Social class and the new towns', *Urban Studies*, **5**, 33–58
HERBERT, D. T. (1968). 'Principal components analysis and British studies of urban social structure', *Prof. Geogr.*, **20**, 280–3
HERBERT, D. T. & WILLIAMS, W. M. (1962). 'Some new techniques for studying urban sub-divisions', *Geogr. Polonica*, **2**, 93–117
HOBSBAWM, E. J. (1964a). 'The nineteenth century London labour market', in *London: Aspects of change*, ed. R. Glass, pp. 7–22 (London)
HOBSBAWM, E. J. (1964b). *Labouring men* (London)
HOLE, W. V. (1965). 'The housing of the working class in Britain, 1850–1914: A study of the development of standards and methods of provision', London University PhD thesis
HOLE, W. V. & POUNTNEY, M. T. (1971). *Trends in population, housing and occupancy rates, 1861–1971*, Building Research Station Research Paper
HOLMES, R. S. (1973). 'Ownership and migration from a study of rate books', *Area*, **5**, 242–50

Bibliography

HOSKINS, W. G. (1957). *Provincial England* (London)
HOUSE, J. W., *Papers on migration and mobility in North-East England*, general editor of studies produced for the Ministry of Labour at the University of Newcastle
HOUSING & CONSTRUCTION STATISTICS, Published quarterly by the Dept of the Environment
HOYT, H. (1939). *The structure and growth of residential neighbourhoods* (Washington DC)
ILLERIS, S. & PEDERSEN, A. (1968). 'Central places and functional regions in Denmark: A factor analysis of telephone traffic', *Lund Studies in Geogr.*, **30b**
INSTITUTE OF MUNICIPAL TREASURERS (1962). *The rating of dwellings: A history and general survey* (London)
JACKSON, A. A. (1973). *Semi-detached London, 1900–39* (London)
JACKSON, J. A. (1968). *Social stratification* (Cambridge)
JACKSON, J. A. (1969). *Migration* (Cambridge, 1969)
JANSEN, C. (1969). 'Some sociological aspects of migration', in *Migration*, ed. J. A. Jackson, pp. 60–73 (Cambridge)
JEPHCOTT, P. (1972). *Homes in high flats* (London)
JOHNS, E. (1965). *British townscapes* (London)
JOHNSON, J. H., SALT, J. & WOOD, P. (1974). *Housing and migration of labour in England and Wales* (London)
JOHNSTON, R. J. (1966). 'The location of high-status residential areas', *Geogr. Annaler*, **48b**, 23–36
JOHNSTON, R. J. (1968). 'An outline of the development of Melbourne's street pattern', *Aus. Geogr.*, **10**, 543–65
JOHNSTON, R. J. (1969*a*). 'Processes of change in the high-status residential areas of Christchurch, 1951–64', *N.Z. Geogr.*, **25**, 650–8
JOHNSTON, R. J. (1969*b*). 'Migration in Greater London', *Trans. Inst. Brit. Geogr.*, **47**, 45–76
JOHNSTON, R. J. (1969*c*). 'Towards an analytic study of the townscape: The residential building fabric', *Geogr. Annaler*, **51b**, 20–32
JOHNSTON, R. J. (1971). *Urban residential patterns* (London)
JONES, E. (1960). *A social geography of Belfast* (London)
JONES, E. (1966). *Towns and cities* (Oxford)
JONES, G. T. & CLARK, C. (1971). 'The demand for housing', *Centre for Environmental Studies UWP 11*
JONES, H. R. (1970). 'Migration to and from Scotland since 1960', *Trans. Inst. Brit. Geogr.*, **49**, 145–59
JONES, P. N. (1967). 'The segregation of immigrant communities in Birmingham, 1961', *Univ. of Hull, Occ. Papers* **7**
JONES, P. N. (1970). 'Some aspects of the changing distribution of coloured immigrants in Birmingham, 1961–66', *Trans. Inst. Brit. Geogr.*, **50**, 199–219
JONES, R. (1962). 'Segregation in urban residential districts: Examples and research problems', *I.G.U. Symposium on urban geography* (Lund) 433–46
JONES, W. G. (1890). *Leicester stockingers, 1860–1890* (Leicester)
KARN, V. A. (1970). 'New town housing surveys: East Kilbride, Aycliffe, Crawley and Stevenage', *Occ. Papers 8, 9, 10 & 11*, Centre for Urban and Regional studies (Birmingham)
KATZ, M. B. (1970). 'Social structure in Hamilton, Ontario' in *Nineteenth century cities*, ed. S. Thernstrom & R. Sennett, pp. 209–44 (Yale)
KEARSLEY, G. W. & SRIVASTAVA, S. R. (1974). 'The spatial evolution of Glasgow's Asian community', *Scot. Geogr. Mag.*, **90**, 110–24
KELLETT, J. R. (1969). *The role of railways in Victorian cities* (London)
KING, L. J. (1969). *Statistical analysis in geography* (New York)
KIRWAN, R. M. & MARTIN, D. B. (1970). 'The economic basis for models of the housing market', *Centre for Environmental Studies WP 62*
KITSON CLARK, G. (1965). *The making of Victorian England* (London)

220

KNIGHTS, P. (1971). *The plain people of Boston, 1830–60*, (Boston)

KUPER, L. (1953). *Living in towns* (London)

LAMPARD, E. C. (1963). 'Urbanisation and social change: On broadening the scope and relevance of urban history', in *The historian and the city*, ed. O. Handlin & J. Burchard, pp. 225–47 (Cambridge, Mass.)

LAMPARD, E. C. (1968–9). 'Review of The Study of Urban History, ed. H. J. Dyos (London, 1968)', in *Urban History Newsletter*, **11**, 20–3

LEE, T. R. (1973). 'Ethnic and social class factors in residential segregation. Some implications for dispersal', *Env. & Planning*, **5**, 477–90

LEES, L. H. (1970). 'Patterns of lower-class life: Irish slum communities in Nineteenth century London' in *Nineteenth century cities*, ed. S. Thernstrom & R. Sennett, pp. 359–85 (Yale)

LEICESTER CITY PLANNING DEPT (1952). *Written analysis: City development plan, 1952*

LESLIE, G. R. & RICHARDSON, A. H. (1961). 'Life-cycle, career-mobility and the decision to move', *Amer. Sociol. Review*, **26**, 894–902

LEWIS, G. J. (1972). 'Leicester: Urban structure and regional relationships', in *Leicester and its region*, ed. N. Pye, pp. 455–80 (Leicester)

LIEBERSON, S. (1963). *Ethnic patterns in American cities* (New York)

LIND, H. L. (1969). 'Internal migration in Britain', in *Migration*, ed. J. A. Jackson, pp. 74–98 (Cambridge)

LIPSEY, R. G. (1963). *An introduction to positive economics* (London)

LOBEL, M. D. (1969). *Historic towns* (London)

MACDONALD, K. & RIDGE, J. M. (1972). 'Social mobility', in *Trends in British society*, ed. A. H. Halsey, pp. 129–47 (London)

MCGREGOR, O. K. (1957). *Divorce in England* (London)

MCKIE, R. (1971). 'Housing and the Whitehall bulldozer', *I.E.A. Hobart Paper* 52

MACMURRAY, T. & SHOULTS, A. L. (1973). 'Market processes and housing policy' *J.R.T.P.I.*, **59**, 187–203

MARBLE, D. F. & NYUSTEN, J. F. (1963). 'An approach to the measurement of community mean information fields', *Pap. & Proc. Reg. Sci. Assoc.*, **9**, 100–9

MARSH, D. C. (1958). *The changing social structure of England & Wales, 1871–1951* (London)

MARTIN, G. H. & NEWMAN, A. N. (1972). 'The evolution of Leicester, 1066–1835', in *Leicester and its region*, ed. N. Pye, pp. 264–87 (Leicester)

MARTIN, J. R. (1845). *Report on the state of Leicester*, Health of Towns Commission

MARTINDALE, D. ed. (1958) *The city* (New York)

MARWICK, A. (1968). *The deluge* (Harmondsworth)

MASSER, I. (1970). 'A test of some models for predicting inter metropolitan movement of population in England and Wales', *Centre for Environmental Studies UWP 9*

MHLG (Ministry of Housing & Local Government) (1970). *Living in a slum* and *Moving out of a slum*, two reports on the redevelopment of St Mary's, Oldham (HMSO)

MISHAN, E. J. (1969). *Cost-benefit analysis* (London)

MITCHELL, G. D. (ed.) (1954). *Neighbourhood and community* (Liverpool)

MOINDROT, C. (1970). 'Movements of the population in the Birmingham region', in *Readings in the sociology of migration*, ed. C. Jansen (London)

MOLOTCH, H. (1969). 'Racial change in a stable community', *Amer. Jour.S ociol.*, **75**, 226–38

MOORE, E. G. (1966). 'Models of migration and the intra-urban case', *Aus. & NZ Jour. of Soc.*, **2**, 100–16

MOORE, E. G. (1966). 'The nature of intra-urban migration and some relevant research strategies', *Proc. Assoc. Amer. Geogr.*, **1**, 113–16

MORRILL, R. L. & PITTS, F. R. (1967). 'Marriage, migration and the mean information field', *Ann. Assoc. Amer. Geogr.*, **57**, 401–2

MORRIS, R. W. & MOGEY, J. (1965). *The sociology of housing: Studies at Berinsfield* (London)

221

Bibliography

MOSER, C. A. & HALL, J. R. (1954). 'The social grading of occupations', in *Social mobility in Britain*, ed. D. V. Glass, pp. 29–50 (London)

MOSER, C. A. & SCOTT, W. (1961). *British towns* (London)

MOUNFIELD, P. R. (1966; 1967). 'The footwear industry of the East Midlands, IV and V', *East Midlands Geogr.*, **25** (June 1966) 8–23, & **27** (June 1967) 154–75

MOUNFIELD, P. R. (1972). 'The foundations of the modern industrial pattern', in *Leicester and its region*, ed. N. Pye, pp. 363–74 (Leicester)

MOUNFIELD, P. R., FAGG, J. J. & GUDGIN, G. H. (1972). 'The modern industrial scene: Changing patterns of manufacturing in the post-war period', in *Leicester and its region*, ed. N. Pye, pp. 375–403 (Leicester)

MURPHY, P. H. (1974). 'Tramways and the Victorian city: The impact of the tramway companies on urban growth with special reference to Leicester', Cambridge University BA Dissertation, 1974

MUSGROVE, F. (1963). *The migratory elite* (London)

MUTH, R. F. (1969). *Cities and housing* (Chicago)

NEDC (National Economic Development Council) (1968). *New homes in the cities* (London)

NEEDLEMAN, L. (1968). 'Rebuilding or renovation: A reply', *Urban Studies*, **6**, 196–209

NEEDLEMAN, L. (1969). 'The comparative economics of improvement and new building', *Urban Studies*, **7**, 86–90

NEVITT, A. A. (1966). *Housing, taxation and subsidies* (London)

NEWTON, M. P. & JEFFREY, J. R. (1957). *Internal migration: Some aspects of population movements within England and Wales*, GRO Studies on medical and population subjects, 5 (HMSO)

NICHOL, J. (1924). *The history of the hosiery trade* (Leicester)

NOCK, O. S. (1963). *The Great Central railway* (London)

OLSEN, D. J. (1973). 'House upon house', in *The Victorian city*, ed. H. J. Dyos & M. Woolf, pp. 320–58 (London)

OSBORNE, R. H. (1954). 'Population concentrations and conurban tendencies in the middle Trent counties', *East Midlands Geogr.*, **1** (December 1943), 30–6

PAHL, R. E. (1967). 'Social structure and spatial structure', *Centre for Environmental Studies WP10*

PAHL, R. E. (1968). *Readings in urban sociology* (London)

PAHL, R. E. (1970). *Patterns of urban life* (London)

PARK, R. E. (1925). 'The urban community as a spatial pattern and a moral order', in *The Urban community*, ed. E. W. Burgess, pp. 3–18 (Chicago)

PARRY-LEWIS, J. (1965). *Building cycles and economic activity* (London)

PARSONS, T. (1954). *Essays in sociological theory* (New York)

PATTERSON, A. T. (1954). *Radical Leicester* (Leicester)

PEEK, R. A. P. & McWHIRR, A. D. (1972). 'Prehistoric and Roman settlement', in *Leicester and its region*, ed. N. Pye, pp. 195–217 (Leicester)

PERLOFF, H. (1973). 'The development of urban economics in the United States', *Urban Studies*, **10**, 289–302

POTTS, G. (1968–9). 'New Walk in the nineteenth century', *Trans. Leics. Archaeol. Soc.*, **44**, 72–87

POTTS, G. (1969). 'The growth of Highfields', unpublished paper presented at the annual conference of the Urban History Group, Birmingham

PRITCHARD, R. M. (1972). 'Intra-urban migration in Leicester, 1860–1965', Cambridge University PhD Thesis

PYE, N. (1972). 'The regional setting', in *Leicester and its Region*, ed. N. Pye, pp. 1–25 (Leicester)

RANGER, W. (1849). *Preliminary enquiry into the health of Leicester*, Report to the General Board of Health

RAVENSTEIN, E. G. (1889). 'The laws of migration', *J. Royal Stat. Soc.*, **52**, 280–92

RAZZELL, P. E. (1970). 'Statistics and English historical sociology', in *The industrial revolution*, ed. R. M. Hartwell, pp. 101–20 (London)

READ, R. (1881). *Modern Leicester* (Leicester)
REDFORD, A. (1926). *Labour migration in the industrial revolution* (Manchester)
REISSMANN, L. (1964). *The urban process* (New York)
REX, J. (1961). *Key problems in sociological theory* (London)
REX, J. & MOORE, R. (1967). *Race, community and conflict* (London)
RICHARDSON, H. W. (1965). *Economic recovery between the wars, 1932–40* (London)
RICHARDSON, H. W. (1971). *Urban economics* (London)
RICHARDSON, H. W., VIPOND, J. & FURBEY, R. A. (1974). 'Dynamic tests of Hoyt's spatial model' *Town Planning Review*, **45**, 401–14
RICHMOND, I. A. (1955). *Roman Britain* (London)
RIDGE, J. M. (1974). *Social mobility in Britain reconsidered* (Oxford)
RILEY, K. (1973). 'An estimate of the age-distribution of the dwelling-stock in Great Britain', *Urban Studies*, **10**, 373–80
ROBSON, B. T. (1966). 'An ecological analysis of the evolution of residential areas in Sunderland', *Urban Studies*, **3**, 120–42
ROBSON, B. T. (1968). 'New techniques in urban analysis', in *Geography at Aberystwyth*, ed. E. G. Bowen, H. Carter & J. A. Taylor, pp. 235–52 (Aberystwyth)
ROBSON, B. T. (1969). *Urban analysis* (Cambridge)
ROBSON, B. T. (1973). *Urban growth* (London)
RODGERS, H. B. (1952). 'Altrincham – a town of the Manchester conurbation', *Town Planning Review*, **23**, 190–202
RODWIN, L. (1961). *Housing and economic progress* (Harvard)
ROLLETT, C. (1972). 'Housing' in *Trends in British society since, 1900* ed. A. H. Halsey, pp. 284–320 (London)
ROSE, J. (1969). *Colour and citizenship* (Oxford)
ROSS, H. L. (1961–2). 'Reasons for moves from the central city to the suburbs', *Social Forces*, **40**, 261–3
ROSSI, P. H. (1955). *Why families move* (New York)
ROSSI, P. H. (1959). 'Comment', *Amer. J. Sociol.*, **45**, 146–9
ROWLEY, G. & TIPPLE, G. (1974). 'Coloured immigrants within the city: An analysis of housing and travel preferences', *Urban Studies*, **11**, 81–90
ROYAL COMMISSION ON THE FRAMEWORK-KNITTING INDUSTRY, 1845
ROYAL COMMISSION ON THE OWNERSHIP OF LAND, 1871–3
SAMUEL, R. (1973). 'Comers and goers', in *The Victorian city*, ed. H. J. Dyos & M. Woolf, pp. 123–60 (London)
SAUL, S. B. (1962). 'Housebuilding in England and Wales, 1890–1914', *Econ. Hist. Review*, **15**, 119–37
SAVILLE, J. (1957). *Rural depopulation in England and Wales, 1871–1951* (London)
SCHNORE, L. F. (1958). 'Social morphology and human ecology', *Amer. J. Sociol.*, **63**, 620–34
SCHNORE, L. F. & DUNCAN, O. D. (1959). 'Cultural, behavioural and ecological perspectives in the study of social behaviour', *Amer. J. Sociol.*, **65**, 132–46
SCHNORE, L. F. (1964). 'Urban structure and suburban selectivity', *Demography*, **1**, 164–76
SCOTT, A. J. (1968). 'A model of spatial decision-making and locational equilibrium', *Trans. Inst. Brit. Geogr.*, **47**, 99–110
SENIOR, M. L. (1973). 'Approaches to residential location modelling: 1. Urban ecological and spatial interaction models: A review', *Env. & Planning*, **5**, 166–97
SEYMOUR, C. (1915). *Electoral reform in England and Wales* (Yale)
SEYMOUR-PRICE, J. (1958). *Building societies: Their origins and history* (London)
SIGSWORTH, E. M. & WILKINSON, R. K. (1967). 'Rebuilding or renovation?', *Urban Studies*, **4**, 109–21
SIMMIE, J. W. (1974). *Citizens in conflict* (London)
SIMMONS, J. (1965–6). 'The making of modern Leicester', *Trans. Leics. Archaeol. Soc.*, **41**, 41–56
SIMMONS, J. ed. (1967). *A walk through Leicester, 1804* by S. Watts

223

Bibliography

SIMMONS, J. W. (1968). 'Changing residence in the city', *Geogr. Rev.*, **58**, 622–51

SJOBERG, G. (1960). *The pre-industrial city: Past and present* (New York)

SJOBERG, G. (1965). 'Theory and research in urban sociology', in *The study of urbanisation*, ed. P. M. Hauser & L. F. Schnore, pp. 347–98 (New York)

SMAILES, A. E. (1953). *The geography of towns* (London)

SMAILES, A. E. (1955). 'Some reflections on the geographical description and analysis of townscapes', *Trans. Inst. Brit. Geogr.*, **21**, 123–40

SMITH, D. M. (1963). 'The British hosiery industry at the middle of the nineteenth century: An historical study in economic geography', *Trans. Inst. Brit. Geogr.*, **32**, 125–42

SMITH, D. M. (1965). *The industrial archaeology of the East Midlands* (Newton Abbot)

SMITH, M. E. H. (1971). *A guide to housing* (London)

SMITH, M. E. H. (1974). *Supplement to a guide to housing* (London)

SMOLENSKY, E., BECKER, S. & MOLOTCH, H. (1968). 'The prisoner's dilemma and ghetto expansion', *Land Econ.*, **44**, 419–30

Social Trends, 1973

STACEY, M. (1960). *Tradition and change: A study of Banbury* (London)

STEDMAN, M. G. (1958). 'Birmingham', *Trans. Inst. Brit. Geogr.*, **24**, 225–38

STEWART, M. L. (1973). 'Markets, choice and urban planning', *Town Planning Review*, **44**, 203–20

SUTCLIFFE, A. (1974). *Multi-storey living: The British working-class experience* (London)

TANSEY, P. A. (1970). 'Housing in Leicester, 1848–1914', Liverpool University BA dissertation, 1970

TARN, J. M. (1961). 'Housing in urban areas, 1840–1914', Cambridge University PhD thesis

TARN, J. M. (1969), 'Housing in Liverpool and Glasgow: The growth of civic responsibility', *Town Planning Review*, **39**, 123–45

TAYLOR, I. C. (1970). 'The court and cellar dwellings: The eighteenth century origin of the Liverpool slum', *Trans. Hist. Soc. Lancs. & Cheshire*, **122**, 186–97

TAYLOR, R. C. (1969). 'Migration and motivation: a study of determinants and types', in *Migration,* ed. J. A. Jackson, pp. 99–133 (Cambridge)

THEODORSON, G. A. (1961). *Studies in human ecology* (Evanston, Ill.)

THERNSTROM, S. (1964). *Poverty and progress* (Cambridge, Mass.)

THERNSTROM, S. & SENNETT, R. (1970). *Nineteenth century cities* (Yale)

THOMPSON, D. (1950). *England in the nineteenth century* (Harmondsworth)

THRASHER, F. M. (1927). *The gang* (Chicago)

THURSTON, H. S. (1953). 'The urban regions of St. Albans', *Trans. Inst. Brit. Geogr.*, **19**, 44–67

TIMMS, D. W. G. (1962). 'Measures of social defectiveness in two British cities', Cambridge PhD thesis

TIMMS, D. W. G. (1971). *The urban mosaic* (Cambridge)

TOOMEY, D. (1970). 'The importance of social networks in working-class areas', *Urban Studies*, **7**, 259–70

TURVEY, R. (1957). *The economics of real property* (London)

VANCE, J. E. (1967). 'Housing the worker: Determinative and cognitive ties in nineteenth century Birmingham', *Econ. Geogr.*, **43**, 95–127

VCH (1958). *Victoria County History of Leicestershire*, IV, 'The city of Leicester', ed. R. A. McKinley (London)

WAKERLEY, A. (1913). *Report to the commissioners of Inland Revenue on the development of the North Evington estate* (Leicester)

WARD, D. (1971). *Cities and immigrants* (Oxford)

WARNER, S. B. (1962). *Streetcar suburbs* (Philadelphia)

WATSON, C. J. (1971). 'Household movement in West Central Scotland', *Occ. Paper* **21**, Centre for urban and regional studies (Birmingham)

WEBER, M. (1948). *Essays in sociology*, ed. H. Gerth & C. Wright Mills (New York)

WELCH, R. (1971). 'Migration in Britain: Data sources and estimation techniques', *Occ. Paper 18*, Centre for urban and regional studies (Birmingham)

224

WELLS, F. A. (1935). *The British hosiery trade* (London)

WEST MIDLANDS CONURBATION HOUSING SURVEY (1971). Ed. R. Welch (Department of the Environment)

WHITBREAD, C. A. & BIRD, K. (1973). 'Rent, surplus and the evaluation of residential environments', *Reg. Studies*, **7**, 193–223

WHITEHAND, J. W. B. (1965). 'Building types as a basis for settlement classification' in *Essays presented to A. A. Miller*, pp. 291–305 (Reading)

WHITEHAND, J. W. B. (1967). 'Fringe belts: A neglected aspect of urban geography', *Trans. Inst. Brit. Geogr.*, **41**, 223–33

WILDING, P. (1973). 'The Housing and Town Planning Act, 1919: A study in the making of a social policy', *J. Social Policy*, **2**, 317–34

WILLMOTT, P. & YOUNG, M. (1957). *Family and kinship in East London* (London)

WILLMOTT, P. & YOUNG, M. (1960). *Family and class in an English suburb* (London)

WILSON, A. (1974). *Urban and regional models in geography and planning* (London)

WILSON, C. (1965). 'Economy and society in late Victorian Britain', *Econ. Hist. Review*, **18**, 183–98

WINDLEY, T. (1917). *Notes on the work of the sanitary committee* (Leicester)

WIRTH, L. (1928). *The ghetto* (Chicago)

WOHL, A. S. (1971). 'Housing in London', in *Working class housing*, ed. S. D. Chapman, pp. 13–54 (Newton Abbot)

WOHL, A. S. (1973). 'Unfit for human habitation', in *The Victorian city*, ed. H. J. Dyos & M. Woolf, pp. 603–24 (London)

WOLPERT, J. (1964). 'The decision process in a spatial context', *Ann. Assoc. Amer. Geogr.*, **54**, 537–59

WOLPERT, J. (1966). 'Migration as an adjustment to environmental stress', *J. Social Issues*, **4**, 92–102

WOOLF, M. (1967). *The housing situation in England & Wales* Social Survey (HMSO)

WRIGHT MILLS, C. W. (1959). *The sociological imagination* (Oxford)

ZORBAUGH, H. (1929). *The gold coast and the slum* (Chicago)

INDEX

Adams, J. S. 23
age-structure
  and frequency of mobility 25, 150–1, 182
  of householders in 1870 63
  of householders in nineteenth century
    and today 110–11
  at marriage 51, 155, 207
  of population in sample areas 100–2
  of spatial structure of city in 1960s 133,
    142, 146–7, 176–8, 189
Agricultural Depression 208
Aldcroft, D. H. 39, 76, 122
Aldgate, London 63
Allan, C. M. 36
Alnwick, Northumberland 8
Alonso, W. 14
American cities
  study of 8–9, 204
  social mobility in 112, 210
Anderson, M. 50
Anson, D. J. 211
Anstey 129
'aristocracy of labour' in nineteenth
  century 49
Armstrong, W. A. 50, 200
Ashworth, W. 6, 36, 38, 48, 206
Asian immigrants, distribution of 177
Aylestone 75, 145, 161, 165, 169, 193
  sample area 93
  vacant houses in 119
  ward 119, 158
Aylestone Park 70, 88
Aylestone Road 82, 93

Barbolet, R. H. 17
Bartlett, J. N. 206
Beaumont Leys council estate 163
Becker, S. 205
Bedford Street, first hosiery factory in 34
Belgrave 93, 129, 145, 167–8
  as high quality middle class area in 1870
    45
  ward 194
Belgrave Gate 58–9
Belgrave Road 58, 82
Bell, C. R. 213

Berry, B. J. L. 212
Beshers, J. M. 9
Best, G. 38, 42
Best, R. H. 3
bid rents, definition of 204
Bird, K. 13
Birstall 117, 129, 168
Blaby 129
Blackfriars Parish 73
Blackpool 72
Boal, F. W. 18
Boards of Guardians 203
Bogue, D. J. 8
Booth, C. 36, 63, 64
Bott, E. 16
Bournemouth 72
Bowley, M. W. 78
Boyce, R. 23, 192
Bracey, H. E. 16
Braunstone 117, 162, 194
  council estate 78, 88, 212
Briggs, A. 214
British Shoe Corporation 113
Brown, L. A. 24
building, definition of in 1911 Census
  208
building cycles in the economy 115–22
building industry
  in early nineteenth century 38–40
  in late nineteenth century 70–1
  in inter-war period 76–7
  response to changes in demand 117
  since 1945 124
  size structure 204
building regulations 38, 69
building societies
  in nineteenth century 39–40, 70
  in inter-war period 76
  since 1945 124
Burgess, E. W. 8, 18
Burnaby, Rev. F. W. 70
business vote in nineteenth century 43
Byatt, I. C. R. 13

Cairncross, A. K. 14, 120
Camberwell, London 63

227